UNDERSTANDING PARROTS – CUES FROM NATURE

Rosemary Low

INSiGNIS
Publications

ISBN 978-0-9531337-9-6

Cover design: Mandy Beekmans and Rosemary Low

Published in the UK by
INSiGNIS PUBLICATIONS
P.O.BOX 100,
MANSFIELD,
NOTTS NG20 9NZ.

Printedin the UK by
Automedia Ltd,
Belton Road Industrial Park,
Loughborough,
Leicestershire LE11 5GU

CONTENTS

ACKNOWLEDGEMENTS

I am indebted to my illustrator, Mandy Beekmans, whose work brings this book alive in a unique and distinctive way. My requirements given in words were translated to the page with imagination, artistry and humour. All her illustrations were commissioned for this book, except those on pages 39 and 47. For these I thank Yvonne van Zeeland for her permission to reproduce them from her thesis *The feather damaging Grey parrot: An analysis of its behaviour and needs.*

My thanks also to Winny Weinbeck who brought Mandy's work to my attention.

Photographs
All photographs taken by and copyright of the author except those on pages 57 (Parrot Welfare Foundation), 116 and 141 (Soledad Diaz), 135 (John Courtney) and 147 (Mark Scrivener and A. Chaloner).

ABOUT THE ARTIST

Mandy Beekmans is a talented Dutch artist. Her passion for parrots and their behaviour made her the perfect illustrator for this book. She graduated in her master's study in behavioural biology in 2013 during which she focused on the sciences of animal behaviour, cognition and welfare. Her aim is a career in animal behaviour research, preferably in a field where she would contribute to the understanding of behavioural needs and welfare. She is involved in writing research grant proposals for a long-term study into foraging and feeding and feather damaging behaviours of Grey Parrots. She volunteers at a Dutch bird sanctuary once a week to gain more knowledge of wild birds. Mandy has kept Budgerigars since she was a teenager. The cage of her three birds is always open. They enjoy chatting with her, her boyfriend and each other as well as perching on the custom made "Budgie tree".

Mandy Beekmans

Arts and crafts have been part of her life since she was very young. She draws human portraits and birds, including fictional and mythological characters. She started drawing digitally in 2010 and is still learning and practising new styles to improve her digital and traditional skills. She believes that the combination of art and science is a unique way to contribute to available knowledge of and education regarding animals in captivity.

FOREWORD

The title of this book says it all, but first you need to banish from your mind any preconceived notions of parrots being *"little people"* or *"the equivalent to a four year old child"* or any similar nonsense. Instead, open your mind to the wild and natural behavioural forces at work and you will soon discover what truly remarkable creatures parrots are, and always have been.

Sharing your home with a parrot is very different from sharing it with a dog or a cat. Those animals, over the generations, have been selectively bred to retain the characteristics that we like and to remove those that we do not like. Yes, they maintain some behaviours that relate back to their wild origins but there is no escaping the fact that a domestic dog is very different from a wolf in the wild.

Parrots, however, are wild animals. There is simply no difference between a parrot in the wild and one in your care; *except* that the former is in a natural environment, the latter is in a very unnatural one. When we place any animal in an unnatural environment it should hardly be a surprise when we see unnatural behaviours occurring.

We cannot create a rainforest, or for that matter any other complex ecosystem, within our homes, but we can do very simple things to improve the welfare of the parrots in our care. The more we understand about the lives of our birds in the wild and the behavioural forces at work all around and within them, the better equipped we shall be to *change our behaviour* to make the lives of our birds better.

I have encountered countless "problem" parrots over the years with all manner of issues ranging from fear of hands to screaming and, of course, aggression. In *every* case the problem has not been with the bird – it has been with the owner. There is no such thing as a "problem parrot", however, there are many, many problem parrot owners. Please do not be one of them. Open your mind, learn from nature and aim to develop a positive relationship with your bird that is based on trust and empathy. Once you can do that the rewards will be immense, for you and, more importantly, for your parrot.

David Woolcock
Curator of Birds, Paradise Park, Cornwall
(Also trainer for the bird show
which emphasises natural behaviours.)

INTRODUCTION

Parrots are among the most beautiful and intriguing creations of nature. No wonder that for centuries humans have wanted to "possess" them! We captured them and put them in cages – but they are still wild creatures. How different is this monotonous, climate-free, unchanging and artificial location from an environment that is pulsating with life and sounds, shade and sunlight and myriad creatures and, above all, trees and other greenery.

Parrots are a very ancient order of birds that evolved probably eighty to ninety million years ago during the Cretaceous period. To me it is desperately sad that since about the beginning of the 20th century man has had such a catastrophic impact on their habitats (deforestation) or the lives of millions of them (trapping) that (as of 2014) 28% of all the 350 parrot species were threatened with extinction. Parrots are the most endangered group of birds in existence.

Many people throughout the world are privileged to live with parrots, to interact with them on a daily basis, yet they know little about their wild relatives. I want them to know. I want them to care what is happening to them and to know how they live in nature. In this book I have tried to weave together details of their lives in the wild and relate them to what they need when kept in our homes and aviaries.

Up until the 1980s only a handful of parrot species had been studied in their natural habitats. Remarkably little was known about their lives, in comparison with most other groups of birds. This is because parrots offer a complex set of difficulties to scientists. They are problematical to catch and only in the smaller species is ringing practical; in recent years electronic tagging has been used. Most species range over large areas (either seasonally or on a daily basis), most nest in cavities high above the ground and they can be very hard to detect in the canopy. Furthermore, many species are endangered and occur in small numbers; even locating them can be a problem.

The number of studies on wild parrot populations is now increasing all the time, with countless scientific papers relating to them, some of them on intriguing or obscure or hitherto unknown aspects. While I have drawn to some degree on this immense wealth of knowledge, the emphasis in this book is on what I have observed over more than half a century of parrots in my care, and parrots in their natural habitats in twenty-seven countries since 1974.

Nothing in my life has compared to the joy that birds have given me – wild birds in my garden, birds in the tropics and, of course, parrots everywhere. I hope that readers of this book will translate some of these pages into ways of bringing joy into the lives of their parrots, and therefore enriching their own experiences – when they realise that it is all about understanding and compassion.

FACING EXINCTION

Many of the species mentioned in this book are in an IUCN* threat category and are prefaced by the words Critically Endangered (simplified definition: 50% risk of extinction within ten years) or Endangered (20% risk in 20 years), Vulnerable (based on reduced population size and/or range), Near-threatened (not yet qualifying for above categories but likely to do so in the near future). No book on parrots would be complete without making its readers aware of this.

* International Union for the Conservation of Nature

PART I.

THE PARTS THAT MAKE UP THE WHOLE

***The power of the beak: feeding hatch destroyed by* Hyacinthine Macaws.**

From a photograph by the author

1. THE BEAK

If asked to define the characteristics that distinguish a species of bird I would list the beak (size and shape), the plumage (colour and perhaps ornamentation), size, and gait (method of walking). In three of these characteristics parrots stand out: unique beak shape, exceptionally colourful plumage and a somewhat pigeon-toed style of walking which is often viewed as comical.

The parrot family is characterised by the curved upper mandible; indeed, in North America the term "hookbill" is often used when referring to this group of birds. The bill shape and the corrugations inside the upper mandible evolved to enable parrots to de-husk seeds, indicating that parrots were derived from a seed-eating specialist. However, some species have, over time, evolved with a different beak shape.

Beak shape and purpose

The cutting edge of the lower mandible has a slight gap towards the centre in most species (but not the nectar-feeding lories), which makes it easier for a nut or a seed to be held in place. This is usually not very pronounced except in the large macaws.

Not all parrots have the strongly curved bill shape. There are several exceptions, in which the bill is elongated and only slightly curved. In both cases closely related species have the upper mandible normally shaped, indicating that the lengthened bill is an adaptation to exploit different food sources. The Long-billed Corella (Slender-billed Cockatoo – *Cacatua tenuirostris*) and the Little Corella *(Cacatua sanguinea)* occur in the same areas and differ little except for the beak shape. They find much of their food by digging in the ground but the former can dig deeper, therefore there is less competition for the same food sources.

Also in Australia, the male of the Red-capped Parrot (known as Pileated Parakeet in European aviculture) has plumage comprised of contrasting areas of brilliant colours such as might be found on an artist's palette – crimson, purple-blue, violet, lime green and dark green. Such rich hues! It has a narrow elongated upper mandible which enables it to remove seeds from the deep capsules of marri *(Eucalyptus calophylla).* Biting off a capsule and holding it in the foot, it looks as though it is eating a bulbous ice-cream cone!

In Chile and Argentina, the Slender-billed Parakeet or Conure *(Enicognathus leptorhynchus)* has an even more elongated and thinner upper mandible. It digs up bulbous roots. Both it and the closely related Austral Conure *(E. ferrugineus),* which has a normally shaped bill, feed on a range of items, including the seeds of *Nothofagus* beech trees and Araucaria ("monkey puzzle" trees). Insect larvae, including the swollen galls on the leaves of beech, are an important part of the diet of the Austral Parakeet.

The range of bill shapes among parrots reflects their food sources and feeding methods. This is

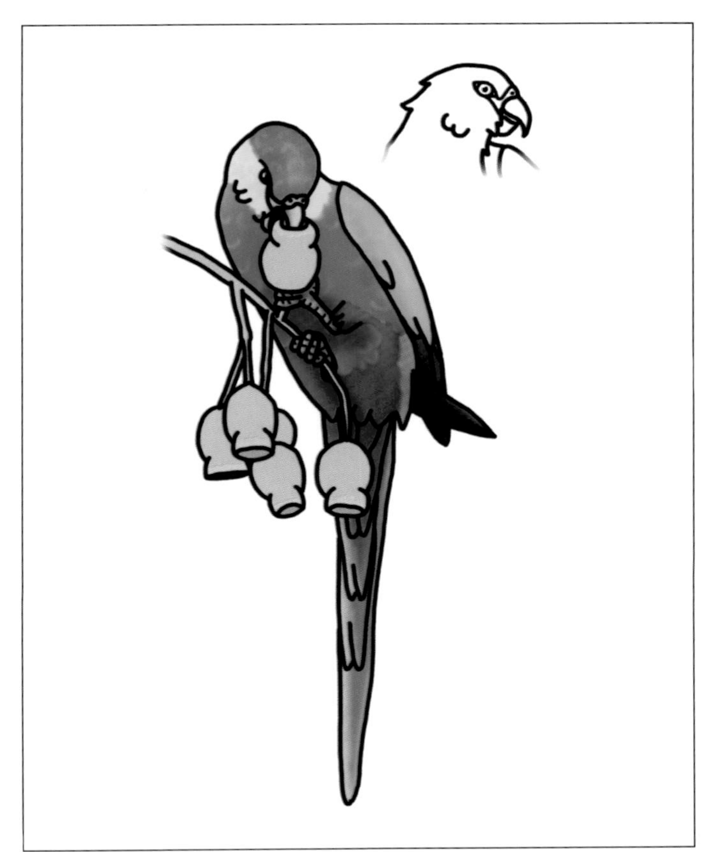

Red-capped Parrot* (Purpureicephalus spurius) *removing seeds from the deep capsules of marri.

especially marked in the cockatoos. In the majestic Palm Cockatoo *(Probosciger aterrimus)* the two mandibles do not fit together. There is a gap in which one can see the red, black-tipped tongue. The beak has two flat projecting areas on the inside of the upper mandible, forming three step-like levels. The large innermost "step" is used for small seed items (including apple pips in captivity), and the middle "step" for larger ones. The outermost step is probably used for tearing at large fruits (Holyoak, 1972).

The different bill structures of closely related species is perfectly demonstrated in the black *Calyptorhynchus* cockatoos, especially in the two white-tailed species. They look almost identical. Baudin's Black Cockatoo *(C. baudinii)* has an elongated upper mandible, which enables it to feed on the seeds of marri, also on insect larvae. The Endangered Carnaby's Black Cockatoo *(C. latirostris)* has a short tip to the upper mandible and eats the smaller seeds of *Pinus, Dryandra* and *Hakea.* Their range overlaps so they do not compete for food sources.

By 1997 Carnaby's Cockatoos had already gone from up to one third of their former breeding sites in the wheatbelt. When I saw them I felt enormously privileged. Black cockatoos are majestic, especially when seen on the wing. When I heard their calls overhead or saw their lazy, flapping flight as a group flew my way, excitement mounted inside me. If they landed nearby, I was enthralled as I watched them feeding, keeping a wary eye on the humans below. I could distinguish their sexes as the male has a black beak whereas that of the female is whitish.

The beaks of different parrot species are worthy of study. The parrot owner needs to know the correct shape for the species kept. The internet is useful here for looking for photos of parrots in the wild; many captive parrots have overgrown bills. Parrots help to keep the cutting surface of the bill in good condition by grinding the mandibles together. This often occurs when they are quietly relaxing.

Even some of the best illustrators incorrectly depict beak length or size. Take the Stella's (Papuan)

Lorikeet *(Charmosyna papou)*, for example. Unlike other members of the genus, the upper mandible has a long, curved tip. In the oft-referred to *Parrots of the World*, William Cooper, one of the best illustrators of parrots, fails to show the elongated tip – no doubt useful when this lorikeet is feeding on the fruits of *Schefflera*. It would be easy for an inexperienced keeper to trim the upper mandible in the belief that it was overgrown.

Beak shape in chicks

In species, such as the Palm Cockatoo and the Slender-billed Conure, with a distinctive beak shape, it can take some weeks or months for young birds to develop an upper mandible which resembles that of the parents. Even in species with a "standard" beak shape this is less curved and/or more blunt in most species.

The chicks of macaws, conures and some other parrots have a pad on each side of the upper mandible, against the edge of the lower mandible. One example is the Red-bellied Macaw *(Orthopsittaca manilata)*. In some species it is visible for only a short time; in others it is more obvious and remains soft until fledging.

In a number of species the lower mandible is comparatively wide in chicks, which probably helps the adults to direct the food more accurately. This trait is, in my experience, especially well developed in the Moluccan Cockatoo *(Cacatua moluccensis)*, which has a projecting lip on each side of the lower mandible.

Bill strength and bites

The strength of bill of parrots is legendary. The largest beak in the parrot world belongs to the Hyacinthine Macaw. From the tip to the edge of the upper mandible measures 8-9cm (about 3½in). That is fearsome! Its destructive powers are immense. The drawing on page one was made

This young Palm Cockatoo, twelve weeks old, will takes some months to develop the beak shape of an adult.

The Moluccan Cockatoo chick has a projecting lip on each side of the lower mandible.

from a photograph I took of the damage meted out to the feeding hatch by a pair of these macaws.

A bite from a parrot has sent many unfortunate people to the Accident and Emergency department of a local hospital. The imprint on human flesh varies with the species. The bill of a *Chalcopsitta* lory is so sharp that it cuts cleanly like a knife. In contrast I have a naughty, nippy "teenage" Crimson-bellied Conure *(Pyrrhura perlata).* A bite from him leaves a bruise and an unsightly blood blister in the same spot. I resorted to wearing the kind of wrist band worn by tennis players when changing his food pots! (There are swing feeders in his enclosure but I cannot use them because *Pyrrhuras* have such fast reactions they exit any gap in a split second!)

One owner of a Moluccan Cockatoo said the bird hated men and would attack them. "When she attacks, she goes for the ear lobe and bites straight through it. We do free ear piercing here", he joked. However, this is no joke as cockatoos and other large parrots can cause very serious injuries.

You would think this might deter thieves – but not if they have had a few drinks. In 2006 a macaw which had been kept in a shop in Somerset for ten years, was stolen. The thief was apprehended because the bird put up a fight. A trail of blood led away from the shop which enabled detectives to recover the thief's DNA. This resulted in the conviction of a man who admitted he had been drinking before the break-in.

Another parrot-mauled man was arrested in Texas. On Christmas Eve 2001, a row took place between two men. One was stabbed and murdered in his home as his cockatoo tried to defend him. It bit the attacker but sadly was killed in the struggle. As the attacker fled from the house, he left blood on the light switch. DNA obtained from it led to the attacker, who admitted he had stabbed the cockatoo with a fork. He was jailed for life. A sad story – but the reporter in the *Guardian* (February 20, 2003) could not resist this crack: "Never argue with the beak"!

The larger parrots need great bill strength because they feed on nuts and seeds which have extremely hard outer shells. Some macaw species can open palm nuts with an especially hard shell that cannot be utilised by other birds. The fruits of the licuri palm *(Syagrus coronata)* should form at least 90% of the diet of the Endangered Lear's Macaw

(Anodorhynchus leari) in north-eastern Brazil – but due to human exploitation of these palms the macaws are frequently forced to feed on corn crops. (See **37. Captive Breeding is not Conservation.**) Each bunch contains hundreds of licuri fruits (nuts), the average length and diameter of each one being 2cm and 1.4cm respectively. The macaws open the nuts by means of perfect transverse cuts.

The skill and speed demonstrated by a macaw when opening palm fruits, is mesmerising to observe. This skill can be put to unusual use in captivity. Steve Martin, bird trainer in the USA, wrote that it was amazing how fast a Blue-throated Macaw *(Ara glaucogularis)* – another palm fruit specialist – could take apart its cage. He described the special screws that hold the cages together as "little more than a mildly challenging enrichment item." All the pairs of these macaws in his show knew how to remove the screws. One pair took out more than 20 screws in less than an hour! (Martin, 2012.)

The owner of a Severe Macaw *(Ara severus)* returned from work one evening to find the macaw had been very busy. The play gym at the top of her roomy cage was tipped at a bizarre angle and the cage door was open. The macaw was perched on a nearby bookcase. She had unscrewed nearly every bolt on the cage and the bolts were scattered around the hall – one lost forever. Toys and perches were discarded at the bottom of the cage; the wing nuts that once held them in place had been tossed across the room. However, nothing in the room had been damaged even although she normally made a beeline for books and TV remote controls. Thereafter, the cage was secured with padlocks when her owner was out!

It was the same story in Australia where the carer of an Eastern Long-billed Corella described its beak as being like a screwdriver. The cockatoo was adept at undoing everything from cage doors to the screws holding the cage together.

What would you do if you were a house-sitter who had accidentally locked herself out of the house in Manchester? While pondering on this problem, she was surprised and embarrassed when a police car arrived. The officer informed her that a 999 call had been received from that address. Assuming that it was a mistake, she went to the back of the house where the owner's Ducorp's Cockatoo *(Cacatua ducorpsi)* was seen standing on the telephone, with the receiver off, pecking at the numbers! Fortunately, a neighbour arrived with a spare set of keys.

Beak colour

Unfortunately for parrot breeders, there are no parrots which can be sexed by the beak colour only because those few in which male and female differ also show sexual dimorphism in their plumage. Apart from the big, black *Calyptorhynchus* cockatoos, the bill being dark grey to black in the male and pale-coloured in the female, the only parrots in Australia in which beak colour differs are Eclectus, *Geoffroyus* and King Parrots *(Alisterus).*

The male has the upper mandible orange and the lower mandible black. Two genera of parrots from Asia have the same orange male beak colour, with both mandibles being black in the female: the Blue-rumped Parrot *(Psittinus cyanurus)* and the

Dismantling the cage: "a mildly challenging enrichment item"!

Psittacula parakeets, such as the Ringneck *(P.krameri).*

What is interesting about these species is that the female is the dominant bird (see **27. Matriarchal Societies**) and the male has to overcome his trepidation, yes, fear of her, to display, and then mate. It makes me wonder whether the red bill colour is especially attractive to the female.

In many parrot species fledglings have the beak lighter or duller in colour, or greyish or pinkish-white instead of the stronger or brighter beak colour of adults. I think this acts to instantly identify a young bird, thus protecting it from non-family members of the same species who might otherwise behave aggressively towards it.

Beak injuries

When I was curator at Loro Parque, Tenerife, my partner and I lived in the grounds. We kept our own parrots on the terrace of our apartment above the park's office. Every day we were visited by some of the free-flying Quaker Parakeets *(Myiopsitta monachus)* in the park.

One pair had a nest in a palm tree about 3m (10ft) above our terrace. From our patio I could watch the dramas there unfold. The nest in the next palm tree was less readily observed but its occupants were instantly recognisable by their beak deformities. "Longbeak", who I think was the female, had an immensely overgrown upper mandible whereas her partner, "Beaky", totally lacked an upper mandible. He appreciated the moistened bread that I put out on the terrace but obviously had survived without any problem before we arrived. I enjoyed watching them carrying on with their lives as though their beak deformities did not exist. Parrots are immensely adaptable creatures. Captive birds can survive with serious disabilities, such as a missing wing or foot, provided that their accommodation is adapted to take this into account.

Sensitivity

A parrot's beak is an extremely sensitive part of its anatomy. This is due to the series of pits in the upper and lower mandibles which contain many touch-sensitive cells. It is important for parrot keepers to know this because if there is an injury to the beak the parrot will be in a lot of pain. If the extreme tip is broken, where there are no blood cells, there will be no pain as this is no different to our finger nails: it does not hurt when we cut them. If the vein near the tip of an overgrown beak is cut, bleeding will occur and much pain will be caused.

This will prevent a parrot from eating anything but the softest food or liquid food. It might lose the will to live. This is why the beak should be trimmed only if absolutely necessary – and by a veterinarian who is skilled in this – most are not. This is not always an easy task and could result in painful and permanent injury.

Research on chickens that had the tip of the beak amputated – to prevent them pecking each other's feathers – revealed something very disturbing that probably also applies to parrots whose beaks are inexpertly trimmed. In chickens, a heated blade simultaneously cuts and cauterises the beak . The initial pain was said to last up to 48 seconds, followed by a pain-free period of several hours. Young chickens apparently suffered less pain but older birds not only appeared to experience more discomfort but – note the length of time – fifty-six weeks later they preened less and made fewer exploratory pecking movements than birds whose beaks had not been trimmed.

A regular supply of branches for gnawing will usually prevent the need for a beak trim. In some captive parrots, such as caiques and fig parrots, the beak will become overgrown quite quickly if branches or other wood is not available most of the time. A slightly overgrown beak should be left alone as it seems likely that regular trimming encourages overgrowth. Instead, the emphasis should be on hard items for gnawing. (See **22. The Need to Gnaw.**) Note that one cause of an overgrown beak is liver disease.

Preening tool

The tip of a parrot's upper mandible is a precision instrument. Being so sensitive, it is the perfect tool for preening itself or preening its companion. The latter is known as allopreening and serves several purposes. An important one is to remove the sheaths on new feathers on areas such as the head which a bird cannot reach itself. Allopreening also strengthens the pair bond.

2. THE TONGUE

The tongue of a parrot is thick and fleshy, thus making it highly dexterous. Compare this with the tongue of another group of primarily fruit-eating birds, the toucan. Its tongue is more like a thread – quite amazingly thin. Contrary to general belief, the tongue of a parrot does play a part in modifying the sounds it makes. (See **16. Mimicry**).

Tactile and muscular, it is a versatile tool that enables it to take advantage of a wide range of foods. It allows a Hyacinthine Macaw to crush nuts on which a hammer makes no impression. The power of the beak is enormous – but without the tongue to position the nut precisely along its seam, the beak would not be as effective. The macaw will also cut a leaf and use this to place against the nut for a better grip. The tongue's versatility was apparent when I saw a Hyacinthine delicately picking the tiny seeds from the outer part of a strawberry – and then discard the pulp.

Many parrot species feed on seeds and nuts on which the outer layer must be removed. Some species can de-husk seeds of many shapes and sizes because their anatomy is adapted for this purpose. The parrot places the seed between the frontal edge of the lower mandible and the groove on the horny plate of the upper mandible. The inner surface has a pattern of corrugations that help to keep the seed in place. The front of the upper mandible penetrates the husk and removes the part against the bill, then the seed is rotated with the tip of the tongue and the rest of the husk is removed. Rapid movements of the tongue increase the speed and efficiency of the process.

The tongue can be modified to a degree to enable different species to take advantage of a particular food source. One of the strangest in the parrot world belongs to the Palm Cockatoo. It is pink with a reinforced black tip. The "reinforced" area is the epithelium, a layer of tightly packed cells that protect the tip. I have often watched these cockatoos eating, which they do with great delicacy. One can observe that the tongue is even more tactile than in most large parrots and can be used with precision.

Cockatoos and lories, unlike most parrots, use their tongues to test items and explore surfaces. The lories and lorikeets, often known as "brush-tongued parrots", have the most highly adapted tongue – a complex instrument, furnished at the tip with "brushes". These are tiny papillae or projections which increase the surface area of the tongue and enable lories to rapidly collect nectar and to scoop up minute particles of pollen from a blossom. Pollen is the main protein source, essential when young are being reared.

If you have never seen a lory's tongue and you meet a tame lory, hold a piece of fruit or a flower just outside its cage or aviary. It will stretch out its tongue and unfurl rows of tiny brushes at the tip. When not feeding, its tongue resembles that of other parrots.

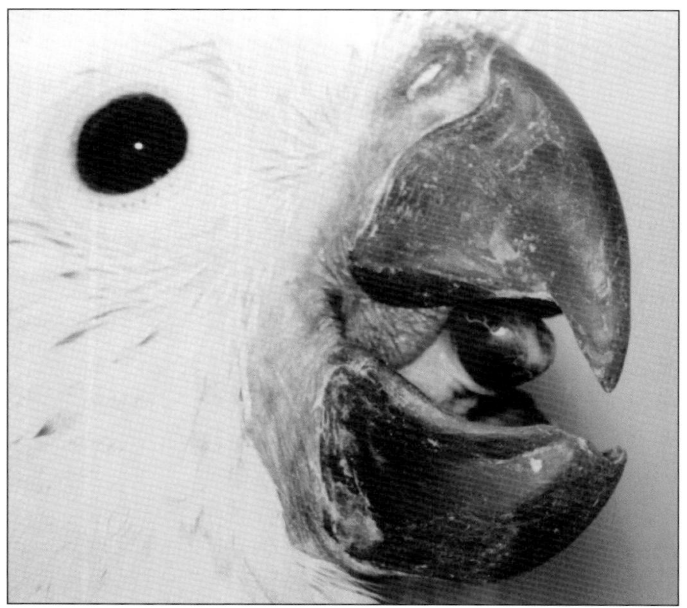

Top left:* *Tongue of normal size and shape: Rose-headed Conure* (Pyrrhura rhodocephala). *Top right: Black tongue, marked with yellow, of Hyacinthine Macaw. Bottom left: Tongue of the Moluccan Cockatoo. Bottom right: The many papillae can clearly be seen on the tongue of this* Eos *lory.

The Swift Parrot *(Lathamus discolor)* is often cited as an example of another parrot whose tongue is adapted to feed on nectar. However, this is much less sophisticated than that of the lorikeets, being shorter, broader, less flexible and containing fine, hair-like fibres. This parrot is not closely related to the lorikeets – but to the Australian broad-tailed parakeets.

The fleshy tongue of a parrot contains numerous blood vessels. An injury to the tongue could be life-threatening, due to blood loss. Every precaution must be taken by keepers to minimise injuries by, for example, double-wiring partitions between aviaries to prevent attacks by neighbouring birds. A tongue injury makes it difficult for the injured bird to feed. It might need to be given fluids to prevent dehydration.

Stitching a tongue injury would be extremely painful. One parrot owner, faced with an emergency and no vet within several hours of travel, gave treatment to a parrot with a bitten tongue in a painless and very effective way. She used human tissue glue! It worked perfectly.

3. THE GORGEOUS EXTERIOR

What is the most gorgeously coloured parrot in existence? This question might have as many as 300 answers! If I ask myself which species made the most impact on me, for its plumage, when I saw it in the wild, I think I would answer: the Collared Lory *(Phigys solitarius).*

Only 20cm (8in) in length, it packs a punch with the power of its colour. No picture can do it justice. Lories are known for their wonderful colours but the intensity and contrasts of this species make it remarkable. Brilliantly plumaged birds often disappear into drabness when seen among foliage but the Collared Lory, high up in the sun, gleams with colour like no other bird I have ever seen. Its unique cape of vivid green seems to stand away from the rest of its plumage.

On two mornings on the Fijian island of Viti Levu it presented a lovely, colourful picture I can never forget – a brilliant red and green lory feeding in a flowering scarlet hibiscus bush. This was not a scene painted by an imaginative artist who enjoyed vivid colours! It was real life!

These lories moved through the leaves with great rapidity. Every now and then a little head appeared and the bright plumage was exposed for a second. The flash of my camera, only 2m (6ft) away, left the lories unperturbed. One stopped still and surveyed me (I was hiding in foliage), unconcerned. They spent much time in the tops of coconut palms, descending lower to search for blossoms, pollen, nectar and insects.

The feathers of this lory were once of extreme importance in Fijian culture. The local name, *Kula,* means red. It also means rich. The red feathers were valued like gold and used as currency. This must have created some difficulties on a windy day!

My dictionary's opening words in the definition of a parrot are "a brightly coloured tropical bird..." – but anyone who knows parrots finds them endearing for reasons that have little or nothing to do with their plumage. Nevertheless, few groups of birds have such an amazing array of feather colours and patterns as the parrots. Almost any species can boast more colours than all the members of one genus of many non-psittacine birds put together.

Some of the colour combinations are remarkable. Every day I marvel at the plumage of my Duivenbode's Lories *(Chalcopsitta duivenbodei).* Where else in nature would you find a colour scheme of rich brown and daffodil yellow, set off with touches of violet-blue? Top fashion designers please note!

Plumage types and colours

Several different types of feathers are found in the plumage of parrots. The most obvious are the contour feathers. They grow from tracts of skin called pterylae; the areas between these tracts (apteria) have no feathers growing from them. This is not apparent unless one parts the plumage because all areas are covered by overlapping feathers. One can, for example, blow the feathers on an Amazon parrot's neck to reveal comparatively large areas which then appear bare. The feather tracts vary in their arrangement in different species. These can be seen most easily in chicks just before the feathers start to erupt.

Parrots also possess down feathers which grow from the skin on all parts of the body. These have a fluffy texture, not tightly knit as are the contour feathers. The colour varies in different species but is generally white, pale yellow or grey. In some Amazons, the newly opened down feathers are yellow but soon fade to white.

A third type of feather in parrots is the powder down – a modified down feather which is believed to grow throughout the life of the bird. The barbs disintegrate to produce a fine powder which is used for cleaning the feathers. It is more evident in some species, notably cockatoos and Grey Parrots, than in others. Cockatoos give off clouds of dust when they shake their plumage. A friend who has a couple of cockatoos in the house says: "If you don't dust every day, you can write your name on the furniture!" This must be borne in mind by those with chest complaints or allergies to bird dust. Spraying parrots daily with warm water helps to reduce the possibly harmful impact to humans.

A few parrots also have bristles (more commonly found in insectivorous birds). These are contour feathers that lack the barbs. In a Lesser Vasa Parrot *(Coracopsis nigra)* that I hand-reared, I noticed bristles around the eyes and cere. Less well developed bristles, and less obvious because of their light colour, can be seen in young Grey Parrots.

Down in chicks

In chicks there are two kinds of down: the first is the natal down (usually white) of the newly hatched which is lost after a few days. The second down (often grey), appears in most species between about ten and twenty days, when the eyes open. It is especially dense in high altitude species or those from cold climates.

Nestling down varies in colour (white, grey or yellow) and density according to the species. Some parrots hatch with only a few wisps of down, others have a moderate covering and some *Calyptorhynchus* cockatoos are marvellous balls of yellow fluff although they are lowland species. However, a few parrot species hatch naked.

I have noticed that chicks reared by their parents have a more profuse down than those which are hand-reared, as seen in the photograph on page 14 of two four-week old parent-reared Grey Parrots. This density could be because of regular preening.

Long white down in a Black-capped Lory* (Lory lorius erythrothorax) *on the day it hatched.

Profuse down of parent-reared Grey Parrot chicks.

Ornamental feathers

There are several types of feather ornaments, that is, feathers that have no practical purpose other than beauty and which might be used in courtship displays. The crest is almost confined to the cockatoos (including the Cockatiel) in which it has various forms and colours. It is interesting that the crest arising from the feathers of the forehead occurs in only one other species of parrot: the Horned Parakeet *(Eunymphicus cornutus)* from New Caledonia. However, unlike the cockatoos, its crest is not erectile. Cockatoo crest formations vary, from curved and conspicuous when not erected and strikingly obvious when displayed, to the crest that lays flat and is hardly apparent until erected, such as Ducorp's Cockatoo.

Most unusual is that formed by the short, wispy feathers of the Gang Gang Cockatoo *(Callocephalon fimbriatum)* – red in the male and grey in the female.

The crest is a truly fascinating feature because, except in the Gang Gang, whose filamentary crest is not erectile, it can transform its wearer's appearance in a split second. It is a barometer of the emotions, indicating a change of mood – especially excitement or aggression. Anyone in close proximity to a raised crest needs to be wary! Cockatoos use their crest to impress or threaten or when they are playing. The unique Hawk-headed Parrot has a totally different kind of crest, formed by elongated nape feathers. These have an element of surprise as they open suddenly to frame the head like a ruff. It is highly effective in instilling fear in the observer! (See page 77).

Tail ornaments in parrots are rare. Racket-tails *(Prioniturus)* are beautiful birds with soft colours and big liquid brown eyes. The several species are unique in that the two central tail feathers have bare shafts and spatulated tips. The Crimson-spotted Racket-tail *(P. flavicans),* at 37cm (14in) the longest member of the genus, has two tail shafts ending in rackets which each measure 8cm (3in) beyond the end of the tail feathers. This includes the spatule which measures 20mm (4/5in). It takes six weeks for one racket feather to grow, after which the second one is moulted and replaced. The bare shaft is wiry and pliable. The preening bird will even bend the long feather over its shoulder.

This is also true of one of my favourite parrots, Stella's Lorikeet. In proportion to its body size it has the longest tail of any parrot. The length, not including the two long feathers, is about 25cm (10in). For fourteen years I have measured the two long moulted tail feathers of my birds. They average about 26cm. However, one of my males consistently exceeds all other Stella's I have kept

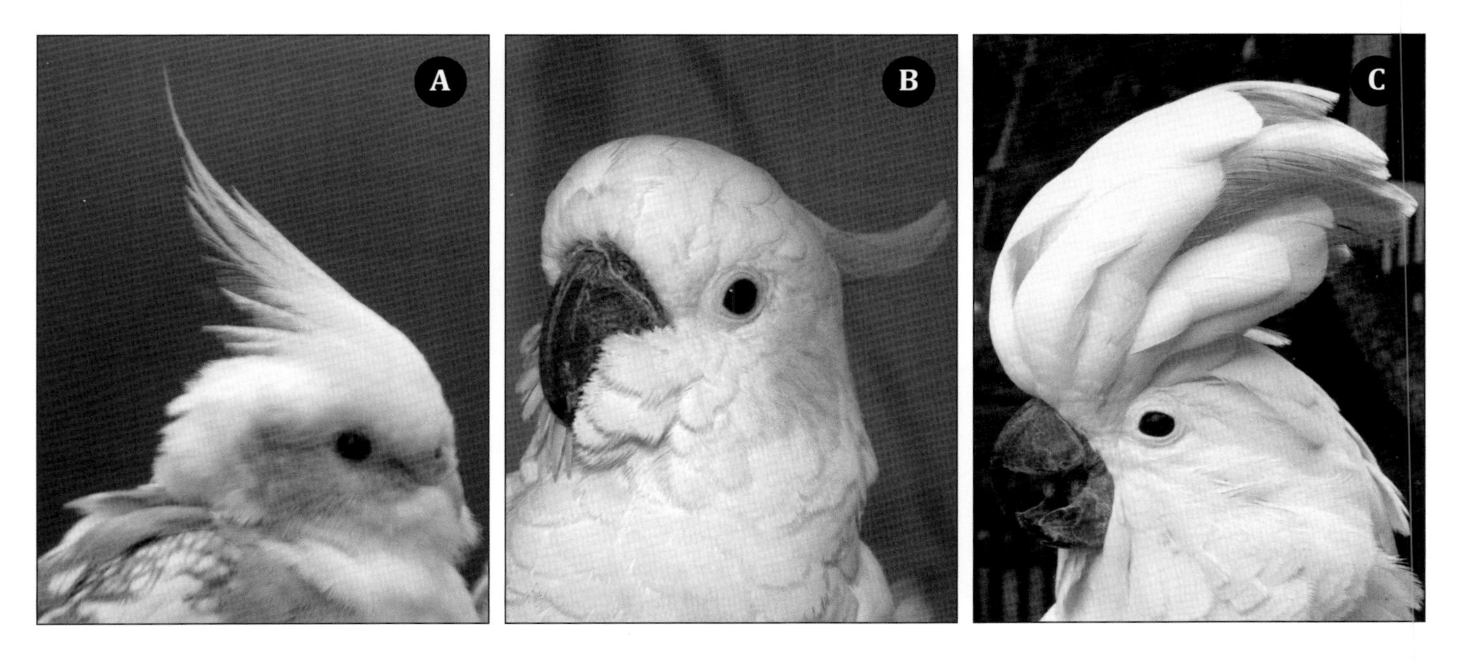

Different forms of crests: A. Cockatiel; B. Citron-crested Cockatoo **(Cacatua sulphurea citrinocristata);** ***C. Moluccan Cockatoo; D. Horned Parakeet; E. Umbrella Cockatoo*** **(Cacatua alba);** ***F. Ducorp's Cockatoo.***

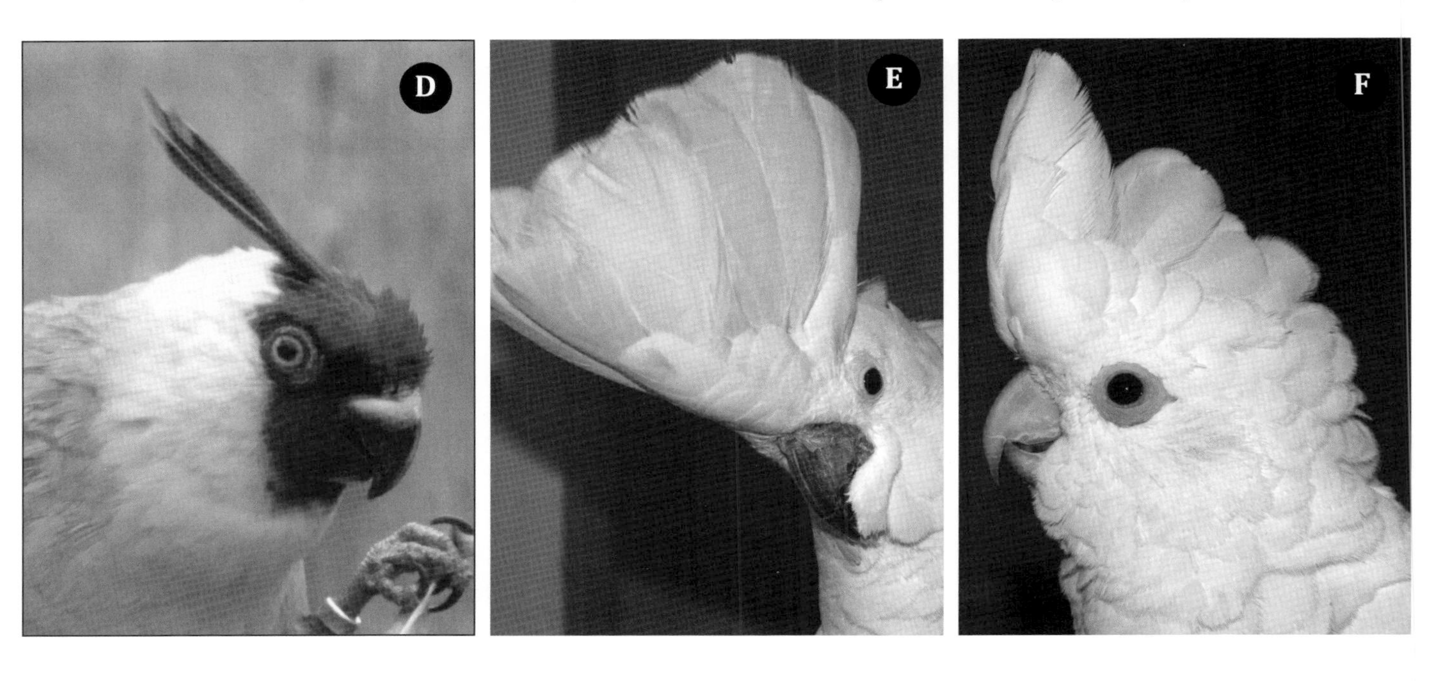

with tail feathers at least 30cm (12in) in length. His longest feather was 33.7cm (just over 13in)! Accuracy can be difficult to achieve as the tip of the feather tends to wear away. It takes about eight weeks for the long feathers to grow and reach their final length.

It is interesting to see how a Stella's preens these long feathers. I have seen my male hold the feather against the perch with his foot, then run his beak through the narrow tip.

Stella's Lorikeets are very active birds and are at their best in large, planted aviaries. (The plants need to be well established as they love chewing greenery.) Only then are you likely to see the way they flick and separate the two long tail feathers. Of course, the very best way to observe them is in the mountains of New Guinea – something which I have been fortunate to experience.

One of the world's most elegant and colourful birds, unlike most lorikeets, it usually flies no higher than canopy level. It is an unforgettable sight to see one or two in flight (they are not flock birds), with the long tail streaming behind. The red females, with the brilliant yellow patch under each green wing, stand out like beacons in flight. The two long tail feathers are yellow for about half their length in the red phase. Unusually, there is a melanistic phase with dark (not necessarily black) plumage in which the tail is grey-green tipped with yellow.

One morning was particularly memorable. Sightings centred on a tall flowering tree. It was attracting lorikeets like bees around a honey pot. The Stella's Lorikeets evoked gasps of delight as their flashes of red and yellow contrasted with the foliage and moss high in the trees. Adults are unusual in having attenuated, slender tips to the primary flight feathers. This is also true of the closely related Josephine's Lorikeet *(Charmosyna josefinae).*

***Primary feather of a Stella's Lorikeet showing attenuated tip, alongside a normally shaped primary feather* (Pyrrhura *conure*).**

Black cockatoos come a close second for tail length and certainly have the broadest tail feathers. The red in the feathers makes them objects of great beauty.

The Princess of Wales Parakeet *(Polytelis alexandrae)* is distinguished by a unique feature. The adult male has an elongated spatule-tip to the third primary in each wing. This might be worn away before the next moult. It resulted in the species being placed in its own genus, *Spathopterus,* in 1895. The spatule was not mentioned in early descriptions and was apparently unnoticed. It was not long before taxonomists rightly rejected the idea that this

Tail feathers shown half life size, from top to bottom: male Red-tailed Black Cockatoo (sub-species* naso*), 26cm long; male Carnaby's (30cm long) and female* naso *(26cm). A one pound coin is shown for size comparison.

feature was important enough to classify the species in a separate genus.

Structural and pigmentary colours

Two types of colours are found in parrot feathers – structural and pigmentary. A feather consists of a central shaft to which the barbs are attached. Each barb has two rows of side branches, the barbules. Hooks on the barbules create an interlocking system. The components responsible for structural colour are present in the barbs and barbules.

Some colours are due mainly to the back-scattering of light from the many keratin cylinders that form the spongy structure of the barbs. (Keratin is the same substance from which the beak is formed, also our finger nails, for example.)

Another factor responsible for feather colour is pigment, such as melanin and carotenoid. The two predominant colours in the feathers of parrots – all 350 plus species – are green and red. Red is found in about 80% of parrot species. Parrots are unusual in that they do not concur with the red, orange and yellow colouring found in plants, fish and other birds, which is produced by chemicals such as carotenoids, called biochromes. The latter are fat-soluble pigments. In parrots, however, a unique set of five molecules is responsible for the red coloration, according to researchers Kevin McGraw and Mary Nogare at Arizona State University. These molecules are called polyneal lipochromes or psittacofulvins.

McGraw found carotenoids in the bodies of parrots but not in the feathers. And the lipochromes found in the feathers were not found anywhere else in the parrots' bodies, implying that they were being manufactured directly at the maturing follicles of the growing plumage. (Lipochrome pigments, such as carotene and lutein, are fat-soluble. They are the natural colouring material of egg yolk, butter and corn.) McGraw said: "The fact that a unique single set of pigments – found nowhere else in the world – is widespread among parrots is an evolutionary novelty." So this is something else that makes parrots unique!

A spectrometer is an instrument used to measure properties of light. Researchers studying Eclectus Parrots in Australia used one to detect the "hidden colour" of males. Rob Heinsohn wrote: "The males spend virtually all their time in the tree-tops and, unlike females, need to blend in with their green surroundings for safety from aerial predators. However, they also need to be bright and showy when they compete for the females' attention at the nest hollows. To achieve this double-act, their green has an extra-quality. It positively glows in ultra-violet wavelengths that are beyond the range of their predators. Males look dull green and camouflaged to hawks and owls (and us) when they are out collecting food, but stunningly gorgeous to other Eclectus back at the nest hollow... The female's red when she slips out of the nest to receive food is one of nature's truly beautiful sights" (Heinsohn, 2012).

The "dull green" when collecting food must have referred to the shade of green perceived when males are in the canopy, out of sunlight, as they never look dull green to humans! Heinsohn found that nests were always located in bright light, thus the females would be able to detect the otherwise "hidden" vibrancy of the green plumage.

White plumage

For the purpose of scientific classification, parrots (Order Psittaciformes) are divided into two families, Cacatuidae and all other parrots, Psittacidae. With the exception of cockatoos, in which half the species have mainly white plumage, white feathers are rare in other parrots. They occur (all in very small areas) only on three lories and in several neotropical parrots such as the Patagonian Conure *(Cyanoliseus patagonus)* and in the breast scalloping of a number of *Pyrrhura* conures and on the head markings of several *Amazona* and *Pionus* parrots. Only in caiques *(Pionites)* which have the entire breast white and in the White-winged Parakeet *(Brotogeris versicolorus)* and the Rose-headed Conure, in which the primaries are white, is white a prominent feature of the plumage.

In Cockatoos the plumage is white because of the composition of the feather barbs, which lack the Dyck texture. This is a structural feature which produces blue and green feathers by the back-scattering of light.

Ultra-violet vision

Colour is one of the many factors that endears parrots to us – but ironically the colours that we see when we look at some species of parrots and some areas of their plumage, are not necessarily the same colours as the parrot perceives (see **5. The Eyes have it!** Ultra-violet Vision). Parrots can discern the colours in ultra-violet wavelengths – unlike humans and, apparently, most other birds. Exposure to ultra-violet light is therefore of extreme importance to captive parrots.

Plumage quality

Scientists believe that the quality of a bird's plumage influences mate selection, as a female instinctively knows that a male with bright undamaged plumage is in good health. She chooses such a bird to pass on her genes.

Scientists in the USA tested the resistance of parrot feathers to feather-degrading bacteria. It was found that white and yellow feathers broke down faster than black, blue, green or red feathers. They believed that some pigments in feathers do not only produce colour but also increase the resistance of feather keratin to bacterial degradation

Veterinarians also know that healthy birds have the best plumage. In the USA two vets, Julie Herbert and Corina Lupu, examined parrot feathers under a microscope in order to evaluate feather magnification as a diagnostic tool. They stated that feather damage results from poor nutrition, compromised digestion and trauma.

Common aberrations noted were blue feathers turning pink, green feathers turning rust or yellow, blue and green feathers developing black patches, grey feathers turning red, and feathers losing colour and intensity. Black discoloration was commonly attributed to liver damage, although the connection between the feather and the liver was not determined. Magnification revealed different types of lesions including those caused by external parasites, mycosis (fungus) and disease.

There is a moral here for breeders keen to buy parrots that show unusual coloration, in the hope of producing a new mutation. Such birds might not be healthy enough to breed. Indeed, they might not live long. On the other hand, they might have a normal life-span. The owner of a White-eared Conure *(Pyrrhura leucotis)* told me that its plumage had changed from green to red to maroon over ten years. Having lived in good health, it seemed unlikely that liver disease was the cause, as is often the case.

Over the years I have seen a number of neotropical parrots, especially Amazons and conures, and even macaws, with abnormal coloration. They show such colours as brownish upper breast and dark red cheeks, wings, lower breast and abdomen. To repeat: this is acquired coloration – not a mutation.

Why are parrots with abnormally coloured plumage rarely seen in the wild? In the case of the occurrence of a mutation – which is not so rare – this is probably because a parrot with unusual coloration acts as a target for a bird of prey aiming at a flock of parrots. It would also be the target of trappers because of the very high prices that a mutation not known in captivity would fetch. Another aspect is that birds of mutations with red eyes, such as lutino (which is known – but rarely – in the wild) have reduced eyesight and are probably less able to avoid predators.

If a parrot of acquired coloration (not a mutation) existed in the wild, this would almost certainly be an indication of ill health and the bird would not survive long.

Plumage types

Not all parrots have the same type of plumage. The feathers of lories and lorikeets tends to be glossier than those of parrots that do not feed primarily on nectar. This is possibly an adaptation to prevent feather soiling by nectar. The *Chalcopsitta* lories have a small bare patch at the side of the lower mandible, perhaps for the same reason.

The moult

Feathers become frayed and worn and must be replaced. The first moult, that of small contour feathers such as the head and breast, occurs at an early age. Small parrots start to moult in their fourth month; according to the size of the species, the moult can commence up to the age of six months. Wing and tail feathers are shed over a period of several weeks, between the ages of six and ten months in medium-sized parrots. Tail feathers are usually the last to be replaced.

Adult birds moult the body feathers annually, over a period of a few weeks. The flight feathers are replaced gradually, for obvious reasons. In the largest parrots, it can take between one and two years for every flight feather to be renewed.

An increase in the protein content of the diet is valuable during the moulting period. Feathers consist of approximately 80% protein!

4. WINGS: THE WONDER OF FLIGHT

Many parrot species are superbly skilful in flight. When you watch them in the wild, you know that they fly not only to move to another location, but for the sheer joy of it. To deprive parrots of this ability, even in the limited space of captivity, is totally unacceptable.

The way parrots fly varies greatly among different genera. Experienced observers often know to which genus a parrot belongs by its flight pattern. However, even in the same genus, there are substantial differences. For me, a most moving aspect of watching Lear's Macaws in their extraordinary habitat of red and yellow sandstone cliffs, was not just in seeing a species nearly lost forever now darkening the dawn sky with sheer numbers, but the extraordinary grace and beauty of its fight. Being familiar with that of the closely related Hyacinthine Macaws (25% heavier), I was unprepared for what I saw.

As the sun crept over the distant horizon, tingeing the sky above with a narrow line of red, the Lear's flew from their roosting cliff in a steady rhythmic flight whose elegance was hypnotising in its beauty. The macaws were like airborne javelins. With their long wings, they are superbly acrobatic in flight, moving up and down the cliff face with consummate ease. Sam Williams (known for his work with Yellow-shouldered Amazons *(Amazona barbadensis)* on Bonaire) spent several weeks at these nesting cliffs. He eloquently described Lear's as "gliding along like a manta ray in the water".

The impressive wingspan of Green-winged Macaws.

To me, large macaws in flight are one of the wonders of nature, with their vibrant colours, impressive size and elegant outlines. Many sightings are imprinted on my memory. In central Brazil the Chapada dos Guimaraes National Park is another landscape of spectacular beauty. A waterfall, at the 800m (2,631ft) summit of a table-top mountain, is a renowned tourist attraction. As I gazed into the canyon below, a pair of Green-winged Macaws *(Ara chloroptera)* flew in complete silence into a large tree on the cliff face, not too far away. Then they departed along the canyon.

The pair flew close to the waterfall, their green and blue wings, skyblue rump and upper tail coverts shining in the sun. An unparalleled vision of beauty! Macaws in flight are always awe-inspiring – and when you can look *down* on them, they engender a feeling of wonder. The power and synchrony of their flight has a unique majesty.

An airborne parrot flock in a coastal setting is a rare and memorable experience. In 2003 I was fortunate to visit the location of the world's largest parrot colony in Argentina. Close to a village called El Cóndor, are cliffs where thousands of Patagonian Conures migrate to nest from other parts of the country. To see dozens of them on the wing, the sun lighting up their colourful plumage, was thrilling. Then every couple of minutes a group of ten or so would rise on the wind above the cliff top, then sink down again to where the multitude of nest entrances decorate the cliff face.

The flight of these parrots is exquisitely graceful; they move through the air like gulls, hanging on the thermals, being carried backwards, then hanging motionless, in a tight little group, their wings almost touching. Parrots riding the thermals! How unexpected was that!

Seeing such huge flocks of parrots might promote the idea that this species is common. Unfortunately its population has crashed by at least 50% since about 1975. The usual reasons apply: trapping for the pet trade, habitat loss and persecution.

Often when you observe parrots flying they are silhouettes; no colours are visible. But their outline aids identification. Amazons have a blunt head-shape and square tails; they do not look aerodynamic. The wing shape tells you a lot about their lifestyle. Parrots with broad, rounded wings, such as Amazons, have relatively slower flight speeds. Their shallow wing beats, mainly below the body, look laboured, as though they must work hard to stay airborne. On a normal day, they do not cover big areas. Most parrots fly with steady wing beats but some *Cacatua* species intersperse wing beats with bouts of gliding.

Some sightings of parrots in flight remain vividly in my memory. Especially when they are unexpected. When visiting the Kakadu National Park, in northern Australia, a track by the Mamukala wetlands passed by an open woodland of pandanus trees and eucalypts. Two thirds of the way along the track the vista changed suddenly as it approached the river. The mudflats were covered in Magpie Geese – thousands of them! Suddenly, ten Red-collared Lorikeets *(Trichoglossus haematodus rubritorquis)* sped by like red arrows, the sun illuminating their fabulous, glorious colours. They truly are jet-propelled skyborne jewels – except no jewel shines so brightly and with such intensity.

The Cockatiel with its long, pointed wings, is also fast and built for covering long distances in search of food. It is a nomadic species, moving around in flocks, following the rain which results in the growth of small seeds – or agricultural crops.

On seeing a species in the wild for the first time the way it flies can be a surprise! Eighteen kilometres off the coast of Fremantle, Western Australia, lies a little island called Rottnest. After I had booked my stay I was told by more than one person that Rock Parrots *(Neophema petrophila)* are now hard to find there. The inference was that in only two days it was not very likely that I would see them.

Hot and extremely arid, the island is characterised by small salt lakes, rocky coastlines and "stacks" (very small islands). With a list of the places said to be visited by Rock Parrots, I searched the island without result for a day, becoming increasingly despondent. Next morning I arose at first light and combed the coastal area around Thompson Bay. Discouraged I headed back to the hotel at 8.30am for breakfast. As I passed the tennis courts near the lighthouse an amazing thing occurred. A Rock Parrot flew right across my path about 3m (10ft) away and perched on the metal piping supporting the wire fence of the tennis court. It was extraordinary the way it happened, as though he was deliberately drawing attention to himself.

He watched me without fear, his yellow breast gleaming in the morning sun. A minute later two more Rock Parrots joined him, each one perched about 60cm (2ft) apart. I moved gradually closer but they showed no inclination to move. They sat still, keeping an eye on me yet quite unconcerned for several minutes, then they flew off. Finally I watched them sweep past in a tight little flock of eight or nine birds, displaying a skill in flight that somehow surprised me. That was another moment to be savoured! I had imagined that they would keep fairly low to the ground and perhaps even show some reluctance to fly. How wrong I was!

If a captive parrot escapes, the species has a bearing on whether it is likely to be reunited with its owner. An escaped Cockatiel, unless extremely tame, is likely to fly away from the area quite quickly, covering many kilometres in a few hours. This can also apply to macaws.

Macaws are not nomadic in the same sense, but often fly long distances to find the foods they need. Some species are specialist feeders, taking the fruits of palm trees as the main part of their diet, so wide areas might be searched for fruiting palms.

Many parrots are extremely strong flyers. This is because they may need to move far, even hundreds of kilometres, between feeding areas at certain times of the year or in response to climatic conditions. A truly migratory species is the Endangered Swift Parakeet which winters on the Australian mainland and returns to coastal Tasmania to breed, usually in August. Its long, pointed wings give it great speed. Its stream-lined body and its wing shape, enable it to turn and veer in flight with no apparent decrease in speed. However, its swiftness results in fatalities, especially in low flight, when collisions with the windows of houses and cars occur.

Flight behaviour in parrots can vary according to weather conditions. Joseph Forshaw studied tiny 14cm (5½in) long Marshall's Fig Parrots *(Cyclopsitta diophthalmus marshalli)* in the Iron Range National Park of Australia's northern tip, Cape York. He observed that in sunny weather they travelled in short flights, sometimes perching to preen and stretch. In overcast or wet conditions they flew straight to the feeding trees. When he imitated their calls as they flew overhead they responded, or circled back to a nearby tree. One male even fluttered down to the lowest branches, only a metre or so above his head! (Forshaw, 1998).

Speed of flight

Galahs have been known to keep pace with a vehicle travelling at 79 kilometres per hour (kmph) or 49 miles per hour (mph) which, if accurate, is fast. In contrast the big black cockatoos have a slow, flapping flight which has been recorded as between 25 and 45kmph (16 to 28 mph). In Bolivia, Red-fronted Macaws *(Ara rubrogenys)* had an estimated normal flight speed of 60kmph (37mph). Amazons have been estimated to fly at approximately 30kmph (19mph). In contrast, the average speed of a House Sparrow is said to be about 18kmph (13mph).

Wing-clipping

Young parrots start to exercise their wings by flapping a few days before they leave the nest. Most people who hand-rear parrots have noticed that at this age they will refuse the first feed in the morning in favour of being allowed to fly around the room. Or at least have a good flap. Their instinct tells them that they must prepare to fly.

Ethics have changed rather slowly during the past 50 years of parrot aviculture, not matching the pace of progress in other spheres of pet keeping. These days docking a dog's tail is illegal – as it should be – but a dog without a tail is only slightly disadvantaged compared with a parrot that cannot fly – a parrot that once could fly but suddenly had this fundamental right denied it. Imagine if you suddenly had a leg amputated. How would that affect your state of mind and your mobility? At least you would understand the reason. For a parrot wing-clipping is a frightening and bewildering experience that leads to serious behavioural and health problems and disadvantages.

Australian parrot behaviour consultant Jim McKendry, with a degree in applied science, wrote: "Simply, parrots are built to behave in a range of specific biologically functional ways. The foundation of that functional behaviour is flight. Indeed, it is when we start to attempt to modify the anatomy of our parrots or create expectations of them that are completely incompatible with the expression of their natural biological tendencies that we experience 'behaviour problems' " (McKendry, 2013).

In his experience of more than a decade of consulting with parrot owners, wing-clipping was probably the primary precursor to many of the most significant behavioural health issues. He wrote: "I don't subscribe to the common thought that wing-clipping is 'a personal choice.' Would that be your bird's choice?

"If we are genuine and authentic about promoting relationships with parrots as pets built on a foundation of respect, trust and appreciation then such decisions should be made based on what is best for the bird – not simply to cater for the limitations of the owner's experience or abilities."

How true that is! Why should a parrot suffer flightlessness, quite possibly for a lifetime, because its "owner" is inexperienced and/or incapable of empathising with its needs?

I believe that most parrots suffer enormous psychological damage from wing-clipping, especially if they were previously able to fly. The natural reaction to a predator or any other danger is flight. If this is denied a parrot, or any bird, it must often experience extreme anxiety. Flight is a survival instinct, especially noticeable in flock species because the reaction is simultaneous.

In addition to the psychological damage, wing-clipping can have a serious impact on physical health: injuries when attempting to fly, infections or wing tumours, obesity, heart disease and shortened life span. It often triggers feather-plucking, especially in Grey Parrots. Full-winged parrots have fewer health problems. If there is a health issue, this is more easily observed than in a sedentary bird that can do little more than perch, climb on the cage bars or walk on the floor. What kind of existence is that?

Clipping the flight feathers of a young, hand-reared parrot can have a devastating impact on its entire future. In the USA, Pamela Clark, a breeder of Grey Parrots, wrote: "Previously I had assumed that it would be necessary to clip wings in order to facilitate the transfer for ownership. To make it easier for the new owner to handle the baby. The 'rules' have always stated that we should clip babies before sending them [to a new] home, if in fact they have learned to fly first."

There are no rules. Compassion and common sense should surely take precedence when dealing with a creature as intelligent and sentient as a parrot, especially a Grey. After one year in which Pamela Clark ceased to wing-clip her young Greys she discovered that they were all "far friendlier and affectionate than those I have clipped. Their freedom of movement allows them to give themselves freely to the human/parrot relationship in a way that is so profound that I will never recover from the impact. The experience allows me to understand that it is not the kindest and most compassionate work to celebrate their fledging and all the skills they develop so enthusiastically during that process, and then take them all away before sending them off to a new home" (Clark, 2002).

It saddens me in the extreme that there are still so many breeders who clip a young parrot's wings before selling it – even without asking the purchaser. Their lack of understanding of the impact it will have on the parrot's development and happiness is almost inconceivable to me. They are mindlessly following other mindless breeders.

One American owner wrote of her Budgerigar *(Melopsittacus undulatus):* "Yes, we allow him to be flighted" as though this was a great privilege. How sad I find it that the superb flying machine that is a Budgerigar, so easy to handle in any household, might be deprived of its powers of flight. People can find reasons (normally invalid) to clip the wings of larger parrots. I doubt very much whether they would do it to a finch or a Canary – so why mutilate a Budgerigar?

A full-winged parrot has much greater confidence and interacts more creatively with everything around it. It can make many more choices in play and in exploring its environment and is thus less dependent on its human companion than a wing-clipped bird which is doomed to long periods of inactivity.

Flight in the home or aviary

Why would anyone want to clip a parrot's wings? The most common reasons are fear of it escaping, fear of it colliding with a window and the desire to have some kind of control over the bird – controlling its movements when it is out of its cage and, apparently, to be able to dominate it. This attitude has no place in the heart or mind of a parrot keeper. As for a parrot flying into a window, this is unlikely if common sense is used. When it is first allowed freedom in the house, it should be taken to a window, several times, and put close to the glass so that its beak is touching it. If it flew against a window once, it would be unlikely to repeat the experience. Unless it was flying extremely fast (seldom possible in an average-sized room) injury would probably not occur.

As for the many parrots that escape through an open door or window, this is sheer carelessness and a mindset that indicates a casual attitude towards keeping a parrot in the home that is totally inappropriate. Lock the doors when the parrot is free and obviously keep the windows closed. Is that so difficult? If it is, your home is not a good place for a parrot to be. Think of the countless escaped parrots that are taken by hawks, die of starvation or perhaps captured to end up in an uninformed home. That which costs nothing, is often treated without care.

I despair when I read of people so certain their parrot is devoted to them that they take it, full-winged and without restraint, into the garden, perched on the shoulder. If a parrot is suddenly frightened, perhaps by a loud noise, it takes off. This is instinctive. Its devotion to a human is irrelevant.

Parrots can also be frightened by unfamiliar objects that appear suddenly. A 20cm (8in) high can of lubricating oil frightened my conures in my aviaries; they gave a screech and took off as one bird. The lories (always with slower reflexes) took off a second later. This innocuous looking blue and black item caused fear simply because they did not know what it was: it was too close. If I had placed the can several feet away, and gradually moved it closer they would not have been afraid.

It is a myth that wing-clipping prevents parrots finding freedom. Many escaped parrots fly out because the carer has not noticed that the bird has moulted its cut flight feathers and grown new ones. Even wing-clipped parrots with a couple of whole flight feathers in each wing can fly but they cannot fly well. They are vulnerable to predation or to being hit by a vehicle. Note that in the UK the list of lost Grey Parrots published in avicultural magazines is longer than that for any other species – and there is a heart-breaking story (as well as an act of carelessness) behind each one.

An inexperienced owner might hesitate to have a full-winged parrot flying around the living room, because of persuading it to return to its cage or to him or her. To train a parrot to fly to your hand, flatten both your hands, palm upwards, one in front of the other. On the hand at the back, place a favourite tit-bit that is not available in its cage, or elsewhere, and call. Use the same call words, such as “Come to me!” each time. When it does fly to you, give it the tit-bit, and/or rub its head and give it a lot of attention. It will soon learn what is required. The treat part of the exercise could gradually be withheld.

Your parrot will associate flying to you as a pleasurable experience, even although it is then put back in its cage on some occasions. This is important. If you always do this when it flies to you, it will be reluctant to do so. Use a little psychology here: also call it when you do not intend to return it to its cage!

This training will prove extremely useful in the unfortunate event that the parrot escapes. However, captive parrots often need to overcome the fear of flying down and this is an important reason why some escaped parrots are not safely recovered. They are not used to flying to the hand from a height and the unfamiliar surroundings of the huge outdoors makes them unable to overcome this fear. As far as is possible in the home, a parrot should be trained to fly to the hand not only in horizontal flight, but from the highest point in the room – a gradual procedure.

An expensive escape
Even if it has been trained to fly down, an escaped parrot might be too frightened to move. In the UK, near Stoke-on-Trent, Cockatiel Georgie, who was used to flying freely around the house, escaped after a door was blown open by the wind. Four days later he was located in a park, at a height of 12m (40ft) in a tree. His owner, Madge Morris, phoned the RSPCA when she could not coax him down. They called the fire service but their ladders were not long enough. Staffordshire Fire and Rescue Service then sent a larger vehicle with a cherry-picker style basket used to rescue people from tall buildings. The Cockatiel must have been even more frightened by the vehicles, people and large crowd of onlookers. No wonder it would not move!

Then Mrs Morris had an idea. She returned to the scene dressed in her pink bathrobe, in which Georgie loved to snuggle. The Cockatiel then flew down to the ground and was captured.

The whole operation was criticised by a member of the TaxPayers’ Alliance for wasting time and cash. Ten fire fighters had spent seven hours trying to coax down the Cockatiel, a police officer had arrived and closed off a road and the RSPCA had also sent an officer. It could only happen in the UK!

The harness
There are many owners who would like to take their parrot outside, in the open air, to literally give it a different perspective on life, but rightly they are not prepared to have its wings clipped. One alternative is the use of a harness. This fits around the body, around the breast and under the wings. There are several sizes made, of course, especially for parrots. As with a dog harness, a lead is attached to it.

When I first heard about harnesses, which were developed in the USA, I was dismayed. This was because the only time that a parrot would have its body enclosed in nature would be grasped in the talons of a bird of prey. Most parrots would instinctively react against wearing a harness. They do not like to be touched, except on the head. Cockatoos are an exception. You will see a pair preening almost any area of its mate's body.

A parrot needs to totally trust the person attempting to put on the harness. Otherwise that person could be badly bitten and the bird would become very agitated. I believe that most parrots find the experience stressful and should not be subjected to this if they show signs of agitation. I recall seeing a harness put on a young Grey Parrot for the first time. Two hours later, when the harness was still in place, it started to pluck its feathers. Sadly this became a habit that was never cured.

Another issue is safety. Some parrots have worked out how to undo the harness. Furthermore, others have actually bitten through the strong fabric used in its construction. It is not unknown for the handler to be lured into a false sense of security and not give the bird his or her full attention. The result is that it has flown off, harness attached.

All the possibilities should be considered. A hand-reared parrot might accept a harness without showing any signs of discomfort or the desire to remove it. In this case its quality of life could be greatly improved with different experiences of the outdoors – the garden or further afield. What I shudder to see is the person who walks, with a harnessed parrot on his shoulder, in crowded shopping centres or other places, where it is subjected to unwelcome attention or frightening experiences. There is only one reason for this: the handler wants to attract attention to himself. Few parrots are so well socialised with people, and amenable to totally unnatural and unknown environments that they can cope with this type of situation.

The worst aspect of the existence of the harness was reported by a manufacturer in Australia. Apparently inexperienced owners believed they could purchase a parrot and harness and leave it tethered outside, unattended. The range of horrific results of such a scenario does not bear thinking about.

Stretching the wings

The design of the traditional parrot cage, which is taller than it is wide, is totally wrong. The most important dimension should be length. Few cages allow a parrot or parakeet to do more than jump from perch to perch. If cages were as wide as they are tall at least the smaller species would be able to fly within the cage. The parrot does not need an expensive cage which looks good within the home. A more practical and suitable cage would be one made from welded mesh, of the appropriate gauge for the species, in which the occupant can actually fly. This would cost probably a quarter of the price of a big fancy cage. A company that works in plastics or metal could make the drawers for the cage bottom and a table to raise the cage off ground level.

Note that in the UK, under Section 8 of the Wildlife and Countryside Act, it is an offence to confine a bird in a cage in which it cannot freely stretch its wings in length, breadth and height when both wings are fully extended.

5. THE EYES HAVE IT!

The expression "The Eyes have it" surfaced in 1945 with an animated short film starring Donald Duck. Whether a duck's vision is any better than a parrot's is debatable but, in my view, the eyes of parrots are more expressive than those of ducks! When you look into the eyes of many parrots you can see that they are sentient, intelligent creatures. Just look – especially at the larger parrots, and you will wonder how anyone can doubt this.

Parrots have excellent vision. They can see distant objects that the human eye cannot detect. I have often seen my aviary parrots looking skywards and emitting alarm calls. On one occasion, two of my Amazons were craning their necks and tilting their heads, making predator alert calls. As usual I could see nothing, but I kept watching the sky. After at least 50 seconds, probably more, a small aircraft came into view. Its silhouette was not unlike that of a hawk.

When a human and a parrot look at the same scene, they do not see exactly the same picture. Because the eyes of parrots are set well to the sides of the head, the visual field is about 300 degrees. This means that without turning their heads they can see almost all the way round them – but need to turn the head slightly to see right behind them. It is said that they can focus each eye independently.

When both eyes see the same image, as in humans and owls, this is called binocular vision. It enables depth (distance) perception to occur. However, this is linked with field of view and varies according to the species and its lifestyle. An owl (like a human), has a relatively small field of view but has binocular vision in most of it. This helps in precisely locating its prey – a mouse on the ground, for example. In contrast, a parrot has a very large field of view. It can see almost all around it except at the back of the head. Less than 20% of its view is binocular.

The large field of view assists in detecting predators. The avian eye can focus much faster than the human eye which is why, if you watch a flock of conures, for example, at the very first hint of danger the whole flock is in the air.

Parrots not only have good eyesight: they are extremely observant. One parrot keeper in Los Angeles dwelt in the top apartment on the slope of a hill. His three parrots lived in a small aviary on the roof of the building. One day he noticed his female Lesser Sulphur-crested Cockatoo *(Cacatua sulphurea)* stretched tall in fear; her screams were piercing the sky. It took him a while to find the source of her panic. A large crane had been erected about one and a half miles away on a construction site. That morning it had swung into action for the first time. It was several days before she perceived it as non-threatening and accepted it as part of the landscape. Talking of threats, this once-common species in the trade is now Critically Endangered – for that very fact, trapping on an excessive scale.

Nictitating membrane

This is the eyelid which is seldom seen by humans but, when it is, appears as a milky film over the eye. Everyone who has photographed birds knows about it. The speed of the camera often captures this membrane, making the photograph unusable! The purpose of this translucent film is to protect the eye without the necessity of closing the outer eyelid. The bird can still see, though not as clearly as when the membrane is not in use. Birds in flight must encounter a lot of dust which could have fatal effects if it entered the eye.

Avoiding collisions in flight

Some parrots, especially tiny Australian lorikeets, including the Little and Varied, also small South American parakeets of the genus *Pyrrhura,* have special skills. They can fly at extraordinary high speeds, even when manoeuvring through vegetation. How do they do this without colliding with each other (if in a group) or with foliage? In 2011 research was carried out at the Vision Centre of the University of Queensland. Professor Mandyam Srinivasan and his colleagues discovered that birds can weave safely through dense forest by allowing their eyes to sense the speed of background flow on both sides. They can then adjust their speed accordingly.

As animals travel forward close objects seem to speed by and those further away apparently travel more slowly. (Anyone who watches landscape from a moving train knows that this is true.) They need to ensure that the background images pass at the same speed in both eyes because if a bird flies closer to obstacles on one side the near eye will perceive things passing faster than in the other eye. This imbalance causes the bird to veer away to even out the speed of image flow in both eyes.

This discovery was made by training Budgerigars to fly along a corridor with walls lined with horizontal or vertical stripes. The Budgies flew faster along the walls with horizontal stripes because the stripes were parallel to the bird's direction of flight. When one wall was lined with vertical stripes and the other with horizontal stripes the birds flew closer to the horizontal stripes.

Crepuscular vision

As a desert species locating water is extremely important for the survival of Bourke's Parakeets *(Neopsephotus bourkii).* Apparently in order to reduce the chance of being predated when it is drinking, this activity occurs before sunrise or after sunset. Observers have seen them come in to drink two hours after darkness has fallen, obviously having registered the location beforehand. Scientists have shown that the eyes of Bourke's Parakeets have more rods in the retina and smaller rod diameter which probably increases their vision in dimmer light. However, this results in a slight loss of colour vision at higher light levels.

Ultra-violet vision

When birds are exposed to ultra-violet light – either through sunlight or through a UV lamp made for captive birds – they are able to see a wider range of colours. This is due to the structure of the eye which possesses four cones – one more than the human eye. For wild parrots, the benefits are (it is believed) that they can detect the degree of ripeness of fruits. However, many fruits are eaten before they are ripe. Researchers in the USA found that fruits that reflect UV light are more conspicuous to birds against green leaves, whether in sun or shade. On Barro Colorado Island in Panama scientists found that birds removed nearly all the fruits from plants in UV light but removed fewer than two-fifths from fruits where this light failed to penetrate.

Another aspect of UV vision is that areas of plumage of some birds, including parrots, fluoresce under UV light. This is invisible to the human eye. It appears that this enables certain species to identify the opposite sex. According to Laura Hirst: "Some parrots flash colour to 'mark' their territory, which if misunderstood by another bird could lead to a nasty fight. If owning and housing multiple birds together, UV lighting could reduce stress and aggressive behaviour as communicative signals are much less likely to be misinterpreted..." (Hirst 2014).

In the 1990s Australian researchers reported that the plumage of parrots glows under fluorescent light. This was first revealed by O.Völkner in 1937 in the *Journal of Ornithology* (volume 85, pages 136-146) but it was several decades before the subject received much interest from other scientists. It was Walter Boles' article "Glowing Parrots" that really attracted the world's press (Boles, 1991).

Boles found that cockatoos and most Rosellas and grass parakeets *(Neophemas)* fluoresce strikingly. Certain conditions are apparently necessary: the parrot must have yellow in its plumage and it must be the "correct" kind of yellow. Some parrots have non-fluorescing yellow or both types of yellow in different parts of the plumage. The forehead of the Budgerigar fluoresces so brightly that it looks like the glowing end of a torch to another Budgerigar.

A fascinating finding was that relating to the Golden or Queen of Bavaria's Conure *(Guaruba guarouba)* which is nearly all bright yellow with some green wing feathers. Under UV light only a square patch on the nape fluoresces yet to the human eye it is uniformly yellow.

In the late 1990s Katherine Arnold from the University of Glasgow visited museums in Sydney and Melbourne. She sat in a dark room and looked at the skins of more than 600 parrots under ultra-violet light. She studied sixty species, noted which feathers were fluorescent and found that those used in courtship displays were the most likely to show this trait. She therefore carried out an experiment. She used sunscreen lotion because it reduces UV absorption, thus dulling the level of fluorescence. She applied this lotion to male and female Budgerigars, using a nearly odourless lotion without commercial additives to ensure that odour did not affect the result. The conclusion was that fluorescence of feathers plays a part in sexual choice (Arnold, 2002).

According to Katherine Arnold, Budgerigars perform courtship displays in the early morning when sunlight contains the highest proportion of UV light – the time when their feathers would glow most brightly. Budgerigar breeders should be aware of this because many breeding pairs are kept in buildings with no natural sunlight and without UV light. Pairs that do not breed might be stimulated to do so given exposure to UV light, when their partners might become more sexually attractive!

The research showed that many parrot species that live in rainforest have not evolved fluorescent pigmentation. This is logical because light levels in rainforests are low. The Moustache Parakeet *(Psittacula alexandri)* is a bird of open woodlands and secondary growth, not of dense forest. In its courtship display the male partially opens his wings and quivers them to emphasise the yellow colour of the wing coverts. Possibly this area is fluorescent.

More recent research related to fifty-one parrot species. Thirty-five (68%) of those examined showed strong evidence of fluorescent plumage being associated with courtship displays. Significantly, 97% had such plumage in areas displayed in courtship and 46% in areas not displayed in courtship (Hausmann *et al,* 2003) .

A species with ultra-violet sensitive vision might have bright plumage to impress a female but this plumage might appear dull to avian predators that do not have this type of vision.

Because humans are not sensitive to ultra-violet light we can see that male *Forpus* parrotlets have blue areas of plumage that are green in the female. However, a study of museum skins using specialised equipment revealed the presence of ultra-violet sexual dimorphism that we cannot see. The blue colour of rump feathers is not purely structural but also has fluorescent pigments. The sexes differed in the intensity and wavelength of their fluorescent emission – apparently the first finding of fluorescent sexual dichromatism in birds (Barreira *et al,* 2012).

To the human eye, three-quarters of all parrots appear to be sexually monomorphic, ie, the plumage is alike in male and female. Researchers examined several plumage areas of thirty Blue-fronted Amazons *(Amazona aestiva)* using multi-angle spectrometry. As a result they were able to predict the gender of the birds with 100% accuracy based on plumage colour characteristics. The most significant areas were the forehead and wing tip – but further research was needed (Santos, Elward and Lumeij, 2009).

We know that there are health benefits for captive birds that are exposed to UV light. According to Laura Hirst: "As well as improved communication, research has demonstrated that birds with access to UV display more active and playful behaviours, mutual grooming and improved flight. This is because UV is responsible for certain metabolic processes occurring within the body which overall gives a bird more energy, motivation and improved orientation." (Hirst, 2014).

Ultra-violet rays are important because they stimulate physiological processes, most importantly that of the manufacture of Vitamin D^3. This is needed for calcium absorption, helping to ensure the female has the calcium needed for her eggshells and for the contractions needed to expel the egg.

It is recommended that full spectrum bulbs that emits ultra-violet wavelengths – manufactured for birds – are placed 31cm to 48cm (12in to 18in) above a bird's cage. They are not effective at a distance of several feet. The advice is that a parrot needs only 20 minutes daily of exposure to UV light in order to manufacture sufficient Vitamin D^3. Avian veterinarian Alan Jones warns that exposure during prolonged periods – several hours a day – can cause eye damage. He has seen cataracts develop as a result.

Dilating the pupil

Parrots have the ability to constrict and dilate the pupil of the eye. This is related to behaviour – aggression and excitement, especially. It is often seen in displaying parrots. Humans should take it as a warning – and not approach a parrot when it is in such an excitable state!

Iris colour

The pupil of the eye is black; the colour of the iris varies according to species and age. In adult parrots, according to the species, it varies being white, brown, yellow, orange or red. In the Red-fronted Conure *(Aratinga wagleri frontata)* it is nearer blue than grey. Some species have a double ring, for example, yellow and brown. In young birds it is always brown or grey – clearly differentiated from the more distinctive iris colour of most species. I believe this might be a protection from agonistic adults: they can see at once that the bird is a juvenile and not competition in any sense. The dark iris gives young parrots an air of innocence and vulnerability and of needing the protection of mature members of their community.

The changing iris colour is the only indication of age in many parrot species: once adult eye colour is acquired the last clue has gone.

Yellow-naped Amazon in relaxed mood. The same bird (right) with the iris dilated and showing much more of the orange part of the iris.

Here are some examples.

Species	immature	adult	age adult iris colour
Grey Parrot	grey	yellow	12 to 18 months
Senegal Parrot	black, grey, then yellow-grey	yellow	12 to 18 months
Jardine's Parrot	greyish	red-brown	12 to 18 months
Blue and Yellow Macaw	grey, pale yellow	yellow-grey,	yellow at 18 months but duller than adult
Hahn's Macaw	grey-brown	orange	reddish-brown by about 5 months
Amazon parrots	grey-brown	orange	about six months

Indicators of health

When looking at a parrot one is immediately drawn to its eyes. They are wide, bright and intelligent. If a parrot is sick this is immediately obvious. The eyes look dull and will be frequently half-closed with the lower eyelid raised. This is a warning that must not be ignored.

6. THE FEET

Unlike most birds, parrots have two toes pointing forward and two backwards, as an aid to climbing. It always amuses me to see how many sculptors and artists have failed to notice this and depict parrots with three toes pointing forwards.

The zygodactyl toe arrangement, as it is called, with toes one and four pointing backwards, enables them to hang upside down from branches, allowing them to eat fruits, seeds or flowers that would otherwise not be accessible. The larger parrots have short, powerful legs which make them efficient climbers. Many parrots species spend a lot of time in the canopy and rarely or never descend to the ground. The zygodactyl arrangement of toes means that cages with vertical bars are not recommended for parrots. They need to climb – not slither!

Many parrot species use the foot as a hand when feeding, moving the foot holding the food to the beak, rather than ingesting food directly from a tree. The majority use the left foot. Studies of the Critically Endangered Puerto Rican Parrots *(Amazona vittata)* indicated that they used the left foot to hold food in 94% of observations. In captive birds of the same species nine were left-footed and five were right-footed (Snyder, Wiley and Kepler, 1987).

In Queensland, Allen Friis watched a pair of Barnard's Parakeets *(Barnardius barnardi)* – called Mallee Ringneck in Australia – feeding on the green fruit of the African boxthorn bush which is a noxious weed. He noticed that they always held the fruit in the right foot. He commented that after observing various cockatoo species in Australia he had never seen one use the right foot for holding food (Friis, 2014). From my own observations of captive birds, I would estimate that about 85% are left-footed.

I have never observed an ambidextrous parrot, ie, one that is equally proficient in using the left or the right foot. If they exist they must be rare, as in humans. Scientists call the right or less foot bias "sidedness". One wrote:

> *"... the more biased the sidedness is (at both individual and species level), the more proficient those individuals are at particular tasks."*

He continued with a statement that I queried as untrue. He admitted that it does apply to some animals but not to parrots:

> *"The more biased parrots are towards using one particular foot (and it doesn't matter whether this is the left or the right) the better they are at solving tricky problems – like how to pull up a food reward dangling from the end of a string" (Birkhead, 2012).*

Parrots can broadly be divided into terrestrial and canopy-feeders. The latter are more adept at holding food in the foot, something which many ground-feeding species never do. Australian parakeets and other ground feeders, move through grasses, for example, taking the seed from ground-level or clinging on to narrow stems. They have no need to hold food in the foot and their feet are more slender, less powerful. Some species, such as Galahs, have short legs, enabling them to more easily collect food from the ground.

Canopy feeders often pluck a fruit or seed case, move a short distance away and eat it held in the foot. They have strong, grasping feet – especially apparent in the white cockatoos. These birds, when caged, are likely to shoot out a foot and grab at someone who is passing or at an object that has been placed too close to the cage.

Cockatoos, Amazons and some other parrots, often use their feet when fighting or playing and will even push away another bird in this manner.

When aviary birds are caught up in a net it is apparent that the feet of the larger species should be avoided almost as much as the beak. They can cling tightly to a hand or net and it can be difficult to persuade them to let go. This is why parrots should be caught in a net in flight, because if they land on welded mesh, it is a difficult task to release them. Just as you get one foot free, the parrot will cling to the wire with the other!

The epithelium, or tightly connected cells that cover the feet of parrots, are robust and not easily damaged. It has always been a source of wonder to me how parakeets and parrots can alight on a cactus without puncturing their little feet! On the island of Bonaire in the Netherlands Antilles (off the coast of northern Venezuela) lives the most beautiful of the fourteen sub-species of the Brown-throated Conure *(Aratinga pertinax xanthogenia)*.

It feeds on cactus. I observed that the parakeets could land repeatedly on them with apparent impunity although it was impossible for them to avoid the treacherous looking spines. Parakeets might be immune to the stiff needle-sharp spines because they weigh only 100g or so, but their neighbours, the Yellow-shouldered Amazons, are three times heavier and apparently equally immune. I discovered that the small soft cactus spines were very difficult to remove if embedded in a finger. Touching a cactus as lightly as possible resulted in a fingerful – so how do parrots in arid landscapes escape injury?

Maintenance

The scale-like skin on a parrot's foot dies at intervals. It is shed or carefully removed when a parrot nibbles its feet to keep them clean. Foot-cleaning is a daily task, along with preening the feathers. In captivity the skin can become very dry – another reason for regular baths or spraying.

Style of walking

The way parrots use their feet to get around varies with the species. Lorikeets can hop or walk but most parrots walk. The larger species, such as cockatoos, have an almost rolling gait which looks quite comical and almost like hard work, whereas smaller parrots can run along the ground at speed. Some species in aviaries can rapidly run up wire mesh without using the beak; it is actually rapid jumping as the two feet are moved simultaneously.

In Bolivia, the Red-fronted Macaw *(Ara rubrogenys)*, an Endangered species, has an unusual lifestyle in its arid montane habitat. These macaws spend at least four hours a day walking through fields, looking for cultivated food because most of their natural habitat has been destroyed for agricultural purposes. They peck and scratch like chickens when foraging. Their gait has been described (unfairly!) as "graceless and heavy, rather similar to vultures." Close study of a dozen foraging birds showed that they took an average

The Red-fronted Macaw finds much of its food by walking on the ground.

of one hundred steps per minute, each step being of about 5cm. This equated to about 300m every hour (Boussekey *et al* 1991).

In 2003 I was fortunate enough to watch these unusual macaws in the Mizque valley. I wrote: "The flock members strutted around the earth for all the world like rooks back home searching a field of cereals, or even cockatoos in Australia, but so unlike most parrots in South America!" (Low, 2009). Their bright colours were a stark contrast to the greyish-brown soil, the scattered grey trees with light green leaves and the tall spiny cacti.

Mountains have always held a fascination for me. To watch lorikeets in the highlands of New Guinea was a long-held dream of mine which came true in 1994. Ambua Lodge is situated at 2,100m (6,900ft) at the edge of then pristine rainforest in the Southern Highlands. It has a commanding position overlooking the village of the same name, the Tari Valley grasslands and range after range of distant mountains, half-shrouded in cloud.

Three small lorikeet species, Goldie's *(Trichoglossus goldiei),* Josephine's and Musschenbroek's *(Neopsittacus musschenbroeki),* were present in a tall flowering tree. It was then that I understood why captive birds of these three species are able to run up and down welded mesh like mice. They negotiate the moss-covered trunks and branches, among which they feed, in the same way.

The smallest feet

Which parrots have the smallest feet? This honour goes to the Pygmy Parrots (genus *Micropsitta)* from the Pacific region. Measuring only 9cm (3½in), they are unique. They seldom perch but spend their time creeping around tree trunks and branches, searching for the lichens and fungus on which they feed. This statement sounds like something from a quiz in which you invent a highly unlikely scenario, then ask true or false? It *is* true! The tarsus is the lower part of the leg, below the bend. In Pygmy Parrots it measures 8mm to 10mm (about one third of an inch).

I was fortunate to see the most colourful species, the Red-breasted Pygmy Parrot *(M. bruijinii)*, the only member of the genus found above 1,000m (3,300ft). Two pairs were moving around a slender trunk, feeding, reminding me of Tree-creepers *(Certhia)*. They have extremely long, slender toes with which to grasp the bark.

Their legs are unlike those of other parrots. Bruce Beehler was trapping birds near Wau to ring them when a pair of Pygmy Parrots flew into his mist net. In *A Naturalist in New Guinea* he wrote: “These two were little gems – the male weighing thirteen grams, the female fourteen… This species is another of the natural wonders of the Papuan montane forest: wee, gentle creatures with legs so short that no band could safely be applied. I released them unmarked ...”

Nails

Parrots have strong, curved toenails which help them to grip the bark of trees. But without their sturdy toenails, their feet would be useless for this purpose. It is the nails that give the more acrobatic species such a good grip, enabling them to hang below branches or even swing from the foot of another bird that is dangling below the perch. So strong are the nails of some of the larger parrots that they can dangle from a tree, or in the case of captive birds, from the roof of cage or aviary – holding on with just one nail. This is quite a feat, perhaps equivalent to a human swinging by one finger!

Fledglings have particularly sharp nails. For several days before they leave the nest, they climb to the nest entrance where they are usually fed by one of the parents. Without such pointed nails it would be difficult for them to climb up from the interior.

Anyone who hand-rears parrots knows how painful those young nails can be when they scratch the skin while being handled. Some breeders cut the nails. This is totally wrong. A young parrot must learn to balance on a perch; cut its nails too short and it will suffer frequent falls on to the cage floor. This makes it nervous and destroys its confidence – a bad start in life. Cutting nails encourages further growth; the answer is to file them with a human nail file.

Great care must be taken when cutting a parrot’s nails – and this should happen only when absolutely necessary for the bird’s welfare. It is recommended that one person holds the bird and another trims the nails. If the bird struggles, causing the nail clippers to slip, the vein could be severed, causing bleeding and much pain. If even one nail bleeds, the process should be abandoned because either the person with the clippers is not competent (even although apparently qualified to do this) or it was an unlucky accident. Whichever is the case, loss of blood from one nail is serious enough.

Sharp nails can be used as weapons if a parrot is grabbed by a predator, or backed into a corner in a nest site. The larger parrots instinctively flip over on to the back and strike out with their feet. This form of defence is also practised by chicks in the nest and is particularly evident in macaws and *Aratinga* conures.

PART II.
PSYCHOLOGY

7. UNDERSTANDING BEHAVIOURS

Throughout Europe and North America, and other countries where parrots are popular as pets and aviary birds, there are thousands of unwanted parrots in so-called rescue centres. This is very often because the people who bought them as "pets" failed to understand the challenges of living with a creature that is not domesticated, even although it has probably been bred in captivity. Not only is it essentially a wild bird, it is highly intelligent, unlike some of the smaller creatures kept as human companions. These facts, combined with their complex social behaviour and potential longevity, make parrots, and especially cockatoos, probably the most difficult of all creatures to look after in our homes.

In the USA, Karen Windsor runs a large parrot sanctuary called Foster Parrots, caring for about 500 parrots (described as "unadoptable") of fifty species. She states: "Living with a parrot requires observational skills, a lively imagination and an overall extended investment of mental energy if you are to provide activities and distractions that will keep your parrot engaged.

"We force our parrots to live in our world, and then we don't understand why they scream, beg, become aggressive, become phobic, feather pluck, self-mutilate."

One of the problems parrots face is the inability of many humans to interpret psittacine behaviour. Often people do not try because they are not focusing on the bird's needs but on what they want the bird to do. So the first step is to try to understand every nuance of a parrot's behaviour. This means thinking about what a parrot does, not dismissing its actions without further thought.

Often its behaviour indicates whether it was parent-reared or hand-reared. Most behavioural problems in companion parrots arise from the fact that they were hand-reared and/or because the human is unable to empathise with them. Despite all the evidence, most breeders fail to realise that by hand-rearing the larger species, especially cockatoos, they are creating birds that are unsuitable as long-term human companions. Because these birds are so long-lived this is a tragedy. Adult parrots may become extremely sensitive to environmental change and are unable to control their behaviour, exhibiting bouts of aggression, general anxiety, and excessive screaming. This is related to hand-rearing.

Avian psychiatry has been described as a critical new field for the care of birds and has now reached out to the world of parrots. "In many cases, these symptoms result from the disruption or diminished quality of parent-young developmental interactions, what is referred to as relational trauma. Symptoms are difficult to eradicate because of the enduring nature of attachment processes that actively shape neuroendocrine systems". This was the message of a paper presented at the 30th Annual Association of Avian Veterinarians Conference by G.A.Bradshaw, J.P.Yenkosky and Eileen McCarthy.

Now that qualified people are highlighting what experienced, caring people have known for years, I would make a plea for the end of the practice of hand-rearing for the pet trade, especially cockatoos.

In March 2014 the UK Press carried the story of a woman who was driving along a road when a vehicle in front of her stopped, and dumped an Umbrella Cockatoo *(Cacatua alba)* in its cage at the side of the road. This act no doubt demonstrates how desperate some people become when they can no longer cope with a screaming, demanding, traumatised cockatoo. The Moluccan and the Umbrella are the worst affected. In this case the cockatoo was lucky. It was immediately rescued by a passing motorist. That was the easy part. To find someone who actually understood a cockatoo's psychological and physical needs, and who could provide them over the long term, would be much, much harder.

At the same time it was announced in the Netherlands that hand-rearing of parrots would become illegal from July 2014. It seems that the Dutch have recognised the misfits that hand-rearing creates. Anyone who doubts this should visit a large parrot rescue centre and see parrots that belong neither in the human world nor that of their own kind. Many such birds must be euthanised when they mutilate their own flesh. These are the saddest parrots imaginable, due to a practice that is driven mainly by financial incentive. Breeders must stop and think what they are creating when they hand-rear the larger parrots. They must consider the psychological needs of these unfortunate birds.

Reading behaviours

Ian Rowley noted how Galahs *(Eolophus roseicapillus)* space themselves out when feeding or roosting so that they are no more than 20cm (8in) from the next cockatoo. "If one bird moves towards another, its motivation will be either friendly or agonistic: if friendly, the move will probably be made with a *Close Contact* call or with an invitation to allopreening; if agonistic the move is usually accompanied by a *Gape,* jab of the bill or an attempt to bite the other bird's foot, but usually the intent is perceived and the victim moves away, maintaining its *Individual Distance"* (Rowley, 1990).

This behaviour applies to many other species of parrots. In a captive situation a parrot is unable to move away from a perceived threat. If its carer wants it to come out of the cage and step up and it does not want to at that time, the individual distance has been breached and the parrot either opens its beak (gapes) in a clear gesture of unwillingness or jabs at the hand. It cannot fly away so if the carer ignores the warning signals, the parrot will bite as its only other option in a cage.

Note that the gape is defined by Rowley as a mildly threatening display, given silently. The bill is partly opened, as if preparing to bite, and pointed at the intruder. The behaviour can be interpreted as "Back off or I will bite." Some humans either do not move away or chose to ignore the warning. They deserve to be bitten!

At midday a parrot might be resting and in the evening it might be tired. If the owner tries to

Back off or I will bite!

persuade it out of the cage and it threatens in this way, the bird should be left alone. Even if it does not bite it will be a grumpy bird that will not co-operate. Consideration must be given to the parrot's state of mind. The person might not always be in good humour and must recognise that the same can be true of his feathered companion!

Threatening behaviours in some parrots are too obvious to miss. In conures and macaws, for example, the bird raises it head feathers and swaggers up and down the perch, with wings slightly lowered, and possibly also making lunging movements.

Preening and trust

It is an invitation to friendship when a parrot angles its head towards another bird (or a human), sending out a clear signal that it wishes to be preened. When pairs are formed, it almost certainly helps to maintain the pair bond. It is also a sort of comfort behaviour after something stressful has occurred. In some birds allopreening results in the removal of parasites such as lice and ticks; whether this occurs in parrots is uncertain.

What is certain is that allopreening, which is usually quickly reciprocated, gives a parrot a great deal of pleasure. Most owners of a companion parrot are able to gently scratch its head, performing the same role as a mate of its own species. Humans can also very gently break the waxy sheaths from new feathers of a single bird but only when these are hard and ready to be removed. Scratching a parrot's head is an effective way of strengthening the human-parrot relationship, for demonstrating affection and for rewarding a parrot for good behaviour.

A parrot which allows its owner to "preen" it close to the eye is demonstrating trust. Likewise, my Amazon parrot, my closest companion for 39½ years, would "preen" around my eyes. I considered this to be a great compliment. I would sit and preen her for minutes on end, just as a compatible pair of Amazons would do.

However, it is important to understand that allopreening rarely occurs in some species in which the pair bond is not strong. Take Eclectus Parrots for example. (See **27. Matriarchal Societies** – to understand why this is so.) Allopreening rarely occurs between male and female, although I have seen males preen their young. This means that most Eclectus do not like to have the head rubbed.

Tilting the head (at the same time ruffling the head feathers) is an invitation to be preened. However,

beware this action in Grey Parrots. They have a crafty tendency to use this behaviour to give the owner a false sense of security, apparently enjoying head rubbing for a few seconds and then suddenly turning and nipping the head-rubber! Be especially wary if the Grey's eyes are not closed during this session!

Cockatoo behaviours

Cockatoo have some behaviours not seen in other parrots. These include making very rapid and continuous movements with the tongue which might continue for a couple of minutes or more. They can also bring forward the cheek feathers to cover much of the lower mandible. Those who know cockatoos well believe these are signs of contentment but they cannot be sure about this!

Bill stropping

A behaviour can be repeated incessantly in a captive parrot under stress. In a small cage a cockatoo or parrot might walk up and down the perch incessantly, throwing its head to one side and perhaps screaming every few seconds. Or it might jig up and down on the perch so that onlookers exclaim: "Oh, look! It's dancing!" This is a common example of misinterpreting behaviour.

In the US, a very experienced breeder of conures and macaws, related what happened while he was away for one week. On no account was his new bird carer to diverge from detailed instructions. On his return it seemed that something had gone wrong. All his birds were very unsettled and the plants had grown straight upwards with huge leaves. A male Scarlet Macaw *(Ara macao)* was frantically rubbing the side of his upper mandible backwards and forwards on the perch. A large area had been worn away to expose the pink under layer. "What has happened here?" asked Howard Voren. "Everything has gone crazy."

He received this reply. "You would go crazy too, if the lights weren't turned off for a week." The instructions to turn the lights out at night had probably seemed too obvious to mention. Everything had suffered as a result – but especially the Scarlet Macaw. It had reacted to the stressful situation by incessant bill rubbing.

In his book on Galahs Ian Rowley states that when chewing at the entrance to a nest hollow, a Galah intersperses chewing the bark around the edge with bouts of rubbing alternate sides of the mandibles on the bare area as though stropping a razor. Various suggestions have been made regarding the cause. But at least in some circumstances this action is natural to parrots in the wild. The fact is that we do not understand all the behaviours that parrots exhibit.

When visiting a friend with a Galah that was allowed its freedom in a large room every evening, she was puzzled by its behaviour. Instead of flying around and exploring the countless ropes, perches and toys, it sat on the back of the sofa, endlessly rubbing the sides of its mandible against the wall. It was not used to visitors. I did not approach the Galah or try to interact with it in any way. When I left, its behaviour returned to normal. My presence, an unfamiliar person, had caused it some agitation.

Experienced parrot keepers realise that they should not allow people who their parrot does not know to handle it or to approach closely. Someone with little experience or intuition can cause considerable stress, especially to a Grey, by asking it to interact with a human who it scarcely knows. One cannot blame the bird if, in those circumstances, it bites.

I have sometimes observed very foolhardy behaviour in people who will go straight up to a parrot they do not know and expect it to step on to their hand. It might do so if it is a sweet-natured Mealy Amazon – but to act like that towards a Grey is just to invite a finger dripping with blood!

Colour indicators – facial skin

Parrots can blush – but not from embarrassment! Not many species have areas of bare skin on the face but in several, of those that do, this area can blush pink or red. This is especially noticeable in the Blue and Yellow Macaw *(Ara ararauna)*, which will blush pink when excited.

In the Palm Cockatoo the intensity of the skin colour on its bare cheeks is an indicator of well-being. In a healthy bird and one on a good diet, the bare skin will be deep red. If it is pale pink, avian veterinary advice should be sought because the cockatoo is almost certainly sick.

8. FEATHER PLUCKING: WHY?

A wild parrot that was feather plucked or denuded its young would be an extremely rare case in the wild – almost non-existent. Unfortunately, in captivity an estimated 10%-15% of the larger parrots acquire this distressing habit. It is frequently encountered in the most intelligent species: Greys, macaws and cockatoos. In these and many other parrots, chicks are plucked in the nest-box – a very common happening.

Feathers begin growing while the bird is in the egg. They start as minute swellings in the skin. A feather forms and grows while its base anchors itself into a follicle in the skin. If a feather-damaging bird removes the follicle, it means that feathers will no longer grow from this point.

Feather plucking is the removal or mutilation of feathers. It usually starts on the breast and/or thighs. (See photograph of Grey Parrot on page 42.) Later the inner thighs and wings and, in Greys, the tail feathers, might also be removed or bitten. This behaviour has similar characteristics to hair-pulling in humans (trichotillomania) and hair-pulling in mice, guinea pigs, rabbits, sheep, dogs and cats.

When this behaviour occurs it is important to consult an avian vet. There are so many possible reasons including disease. This can be detected but if the reason is psychological (as it usually is) improving the circumstances under which the parrot lives is essential – and this includes environmental enrichment. Rapid action is advisable because when feather plucking becomes a habit, it is usually impossible to cure. In many birds it becomes an addiction.

A parrot might pluck its companion – as well as itself. Note that if parrots are plucked on the head this is usually the case. However, cockatoos sometimes actually pull out their long crest feathers as these are easily reached.

Many parrot owners fail to realise that when a bird is seen plucking, they should quietly leave the room without another glance at it. If a person shouts "Stop that!" the bird interprets this as a reward of attention. It will pluck even more, not less.

An obsessive-compulsive disorder

In the US medical doctor Stewart Metz, whose love of the Moluccan Cockatoo took him to Seram, founded the Indonesian Parrot Project, focussing on the endangered island cockatoos. As well as his conservation work, he had a great interest in the causes of neurotic feather-plucking and self mutilation in parrots. This is considered by many to be a form of obsessive-compulsive disorder (OCD), similar to flank sucking in dogs, and compulsive self-grooming leading to fur loss. He wrote: "As the OCD persists, it becomes less reversible. This may be due to the actual loss of genes and brain cells in critical areas which control obsessive-compulsive behaviour and/or its brain chemistry...Once the syndrome becomes 'fixed', there is always a risk of recurrence, as in human alcoholism."

It has been suggested that the actual cause is often "exposure to suboptimal environmental conditions in which the animal is faced with unresolvable conflicts" (Dodman et al,1997). "When the source of the conflict is removed, the behaviours may continue to be performed repetitively and pointlessly without a stimulus."

Stewart Metz believed that captivity, especially when complicated by loss of such important aspects as flock structure, flight and food foraging, can lead to feather plucking. This might be reversed by suboptimal concentrations of serotonin (or other neurotransmitters – chemicals that carry messages between different nerve cells) in critical regions of the brain. Plucking and other disorders, especially in young parrots, may be reversible but over time there could be changes in the brain anatomy, leading to little reversibility.

Socialisation is critically important

A study at Rockefeller University in the USA found that Zebra Finches housed in a large colony developed 30% more nerve cells than those housed in isolation. In other words, social interaction can increase the size of the brain. Stewart Metz commented that the absence of adequate socialisation during the commercial hand-rearing of chicks might be a major predisposing factor for later OCD (Metz, 2002) – and therefore feather plucking.

I have no doubt this is true and also believe that the negative effects of hand-rearing can be largely neutralised by placing hand-reared parrots with other young or young adults of their own species immediately they are independent. They need to spend several months in such a situation. If they do not know the behaviour of their own species, they can become hopelessly imprinted on humans. Frustrated by the limitations of such a relationship, excessive preening, then feather plucking is often the result.

Poor socialisation and absence of parents during the rearing period (resulting in failure to learn appropriate preening behaviours) almost certainly makes a parrot more susceptible to feather plucking. It also occurs due to loneliness or boredom in a barren environment without another parrot with whom to interact. Parrots are highly social creatures and bird keepers need to acknowledge that there are limitations to any species' abilities to adapt. For many parrots, living alone is almost intolerable.

Valuable research

Limited formal research had been carried out on the common and serious problems of feather damaging behaviour. Dutch vet Yvonne van Zeeland made it the subject of her PhD thesis, published under the title of *The feather damaging Grey parrot: An analysis of its behaviour and needs.* Unlikely to be encountered by many parrot owners, I believe this thesis should be shared as widely as possible.

The author, a veterinarian and certified parrot behaviour consultant, is a qualified European Specialist in Zoological Medicine (Avian). She states: "Although the consequences of this self-inflicted feather damage may be solely aesthetic, medical issues may also arise due to alterations to the birds' thermoregulatory abilities and

metabolic demands, hemorrhage and/or secondary infections." Some parrots with feather damaging behaviour (FDB) show signs of excessive fear or stress. Fear, phobias, or panic might be displaced and translated into feather plucking. The outcome is often that the bird is given to a parrot sanctuary or it is euthanised.

FDB can be classified as among the abnormal repetitive behaviours which often develop in captive animals due to stress or aversive stimuli and/or the inability to perform species-specific behaviours. Parrots are particularly susceptible due to their intelligence (high-level cognitive abilities) and due to the relatively limited time span during which they have been bred in captivity. Because of these characteristics, life in captivity is inadequate to meet the parrots' social, cognitive and behavioural needs.

Eventually the behaviour may become ritualized and develop from a so-called maladaptive behaviour (trying to adapt to an inappropriate environment) to malfunctional behaviour (with abnormal brain function) – similar to addictions. This might, in part explain, why many feather-plucking birds do not respond to treatment.

Hormonal influence

Feather plucking often develops at the onset of sexual maturity, suggesting hormonal influence. Cyclic or seasonal changes in the extent of FDB may occur, possibly associated with the ambient temperature and humidity, or hormonal changes during the mating season.

A friend of mine has seven Grey Parrots which live a life which is as close to the captive ideal as can be imagined. They co-exist as a small flock but in separate cages, except for one male and female who are housed together. Every day, weather permitting, they spend several hours in a large planted aviary, and are brought indoors for the night. Their diet is excellent and they receive a calcium additive twice weekly. Their cages are spotlessly clean. One May, the female, aged about 20 years, who is kept with the male, appeared to be biting at her wing feathers.

The only factor missing in her life was the possibility to breed. In May the weather becomes warmer and the hours of daylight longer. All the wild birds are breeding; it is only natural that this female desired to do the same. I concluded that the feather biting was the result of this thwarted desire. Many parrots, most often females and Greys, when kept alone or without the facilities to breed, will pluck their breast feathers. Biting at the wing feathers is not common and the only other possible explanation was an infection – but no infection was apparent.

Diets could be to blame. Phytoestrogens are plant-derived compounds found in a wide variety of foods, notably soy, also corn and gorbanzo beans. They can heighten normal sexual activity into sexual frustration, especially in parrots kept without a mate. Hormone treatment for feather plucking is sometimes advised by vets. Drugs, such as Lupron, have been used, to reduce the level of hormone production and to potentially stop feather plucking, but this might only be effective on a temporary basis – and could have side effects.

Environmental enrichment

The value and efficacy of environmental enrichment for Grey Parrots was the subject of various trials by Yvonne van Zeeland which indicated that FDB parrots spent approximately 21% of feeding time foraging while in healthy birds this was approximately 50% (see **34. Foraging and Food**). The results underlined that in feather-damaging Greys the motivation for foraging toys is less, probably because their behaviour has become ritualised and feather damaging has become an addiction. The trials highlight the fact that to try to prevent the onset of FDB in parrots environmental enrichment is extremely important and should be provided for parrots from the earliest possible age.

Other causes

Feather plucking has many causes unrelated to lack of environmental enrichment and hormonal disturbances. They include allergies, injury, wing-clipping, aspergillosis and other diseases. Heavy metal poisoning, including zinc toxicity, is another cause. Lead is seldom used in household items these days, but zinc is common, not only in the galvanising used on wire mesh. It is even found in some rubber products, so parrot toys containing rubber could be suspect. Cages and toys should contain less than 0.5% zinc but it is probable that many contain more.

Stress is a major cause of feather plucking. See **10. Stress and Fear.**

9. ALONE

A wild parrot does not know what it is like to be alone – except one who is incubating eggs. However, this is for a limited period and it will be visited several times daily by its partner. For most of the year parrots exist in family groups or in flocks. Highly social, they are used to – and need – the almost constant presence of their own species. If you are watching parrots and you see a single bird, it is probably because its partner is hidden from view. Or it is a sentinel, guarding its flock, perhaps on the top of a tree.

When you see a pair of macaws or Amazons in flight, they fly almost wing-tip to wing-tip. They need close contact at all times. In contrast, if you watch toucans, for example, you see that they fly in a line – not side by side. At rest, many parrots are in close body contact with another of their species. But in captivity many companion parrots never know this close comforting presence. Worse still, they are left alone for hours at a time.

One newspaper reader commented on a woman who asked for advice because her dog barked all day when left alone. The reader commented: "Animal welfare organisations such as Battersea Dogs' Home recommend leaving a dog for no more than four hours at a time. It is cruel to leave a dog longer than this, and completely contrary to its nature as a pack animal."

Exactly the same can be said for parrots, substituting the word "flock" for "pack". In 2013 the polling firm YouGov surveyed more than 2,000 pet owners for the PDSA. It found that since 2011 the number of dogs regularly left alone for more than five hours rose from 18% to 25%. As the same conditions are likely to apply in homes where parrots are kept, this is a disturbing figure, because so many, perhaps the majority of, companion parrots are single birds.

Should we be surprised that they resort to feather destructive behaviour or become extremely anxious and nippy? The life they lead is totally contrary to that for which they have evolved. So what can be done to alleviate loneliness and boredom? Perhaps one might address the problem from the outset and buy two birds. However, one needs to think ahead regarding the decision to breed. And according to the species, two females might not be compatible.

The other alternative is to acquire a parrot of another species that occupies a different cage, close by. This will help to dispel loneliness and provide interest for both in watching what the other one is doing. Quite often two parrots of different species will communicate with each other and mimic any human language they learn. If they form a strong bond, their attachment to humans might be lessened – but the parrots' interests should come first.

Let's be realistic. It is not right to keep a single parrot that must spend a large part of its day alone, five days (or more) of the week. This is an important point to ponder before the purchase of a parrot is considered.

One lady told me: "I have just bought a copy of your book *The Loving Care of Pet Parrots.* How I wish that I had read it before deciding to buy a cockatoo! Everything you said has come true…!"

In the book I warned about the demands of tame cockatoos that cannot be met by those who go out to work. She was not an unthinking lady who had carelessly embarked on buying a young hand-reared bird. She had wanted to do everything possible for her companion, who she loved dearly. She had even built an aviary in the garden so that when she was out her cockatoo could play and exercise there. The problem was that she worked and when she returned home the cockatoo screamed so much that, very reluctantly, she had to part with it. In fact, the breeder took it back. So already this young bird had lost its first home. I believe that the majority of white cockatoos spend less than three years at their first location.

In Switzerland, the 1981 Animal Protection Law stated that sociable animals, primates, cats and dogs, must be kept with members of their own species. In 2008 the law became even stricter to include any animal that is considered to be sociable, including birds and guinea pigs. Is it not time that other countries adopted this legislation?

In the UK Neil Forbes is one of the few Recognised Specialists in Avian Medicine. He is renowned for his expertise and for his humane approach to bird keeping. He wrote that he would be "in favour of discouraging if not banning the keeping of single pet birds in cages."

He reasoned: "One need only watch and listen to two parrots in a cage, to appreciate how much stimulation they get from each other, to wonder how single caged pet parrot ownership can be recommended. After all, psittacine birds are gregarious, busy and intelligent individuals. Should "would-be owners" complete some form of "online training course", if necessary with a randomised multiple choice questionnaire at the end?" he asked. This would put off impulse purchases and help to ensure new owners were committed and well-informed, prior to ownership" (Forbes, 2014).

The message of this chapter is clear: if no family member is at home during the day, do not contemplate acquiring a parrot that will be alone.

10. STRESS AND FEAR

It seems unlikely that parrots in the wild suffer from stress as known to captive parrots since they can fly away from dangerous or unknown objects. They would, however, be physiologically rather than psychologically stressed by lack of food and water. Predators are, of course, the biggest source of fear.

Captive parrots need a stress-free environment and sympathetic carers. The problem is that many people who look after parrots do not understand what causes stress and fear. Often they are the cause due to the use of negative reinforcement. Pet bird owners often use coercion. Birds should be *asked* to do things, says Greg Glendell, a parrot behaviour consultant, and rewarded when they accept our request; this is the basis of positive reinforcement. This means giving a reward such as food or a head scratch for agreeing to a request. Negative reinforcement is where aversive methods are used on a bird *until* it obeys, such as chasing it around until it steps up.

Stress is often caused by ongoing circumstances, such as a dominant and/or aggressive companion, a cage that is too small, inability to fly, even short periods without food, incorrect diet, large or aggressive birds in adjoining aviaries and hawks near the aviaries. A possible cause of distress is overgrown nails, especially those which have curled round, which get caught in welded mesh.

Parrots suffer deeply and over the long term when their closest companion, be it human or avian, is removed. A temporary separation can be extremely stressful. A permanent one, caused by death, can leave a parrot depressed and intractable. It can take a long time to accept a new person or mate.

I once watched visitors to a bird park laughing at the behaviour of a Moluccan Cockatoo which repeatedly bobbed up and down on the perch, shrieking and opening its crest. It continued to do this as long as the visitors stood in front of its aviary. They interpreted its behaviour as "dancing". I interpreted it as a hand-reared bird desperate for human attention. I suspect this cockatoo was in a continual state of anxiety because it had been hand-reared to be with people and now close human contact had been removed. It is not unusual for former pet parrots to be donated to bird parks – but if they have been hand-reared this can be a very stressful experience for many of them.

There are much more subtle causes of stress which can easily go unnoticed. It can be stressful for a bird to live close to a television set or a computer screen on which there are constantly moving images. The flashing pictures from both create a confusing and disturbing environment. This is especially the case where the television is on all day and late into the night. Lack of sleep will add to a parrot's confusion.

If a parrot lived in a cage that was bare of everything but perches it could feel nervous and exposed.

Hanging several toys in a line, almost like a small curtain, would help to alleviate that. However, these days most parrot owners are well aware of the importance of toys. Even wild-caught birds, that may show little inclination to play with them, will benefit from not living in a bare space. Of course the best way to fill that space is with fresh-cut leafy branches. What could be more natural?

A particularly sad case of stress related to a Severe Macaw biting flesh from its abdomen, resulting in a gaping hole. The owner of the rescue facility where I saw him wanted to uncover the macaw's history for clues to why he had mutilated himself so badly. What she found out was very disturbing. In the first home that could be traced the macaw had suffered the terrible experience of a car crashing into its aviary. In its second home a woman hit the macaw with a long stick and threatened him with the vacuum cleaner. In the third home he was kept in the house with other parrots, notably a Grey Parrot of whom he was afraid. With a lot of loving care and the application of bee propolis to the wound, it eventually healed.

When you care for a parrot, you need to try to put yourself in its place to understand that your own behaviour towards it could be causing it considerable discomfort. As long-time macaw breeder Kashmir Csaky said: "A person may approach a bird in a cage or on a play-stand without thinking about what the bird sees. He approaches quickly, leans over so that his enormous face is close to the bird and then thrusts a huge hand at the bird while demanding, "Step up!" How terrifying for the bird, especially if the person is a stranger. Movement, speed and proximity, all affect how a bird will react to us. We should approach birds slowly, cheerfully, with smooth and confident movements, and watch the bird's body language for any indication that our proximity is stressful, neutral or desired" (Csaky, 2014).

Is she too close, too loud or too threatening?

For someone who is sensitive to the reactions of a parrot, it can be quite painful to watch people who are not used to their presence. When someone suddenly flings out their arms, points or makes jerky movements an aviary bird will usually fly to the furthest perch. A parrot in a cage might crouch in fear or even fall off its perch. Even very tame parrots react nervously. Designed to take flight at the first sign of danger, real or imagined, birds do not wait to find out if their fear is misplaced. If they did, they would soon be in the talons or jaws of a predator. They therefore react instantly to something which makes them nervous.

I know people who like birds – indeed, they even keep birds – but they are too excitable in their behaviour and actions to inspire confidence in a parrot. Instead they create fear – a temporary emotion that will cease soon after that person moves away. A quiet and calm demeanour is what is required and if this is not inherent in a person, it can be difficult to acquire. So, yes, unfortunately, some people cause a parrot stress just by being near it.

Other actions that invoke stress and fear are:

- Standing too close to the cage, especially if placing a finger inside it.
- Wearing a hat or large gloves.
- Wearing bright, solid colours – especially red.
- Placing a large or unfamiliar object close to the cage, or carrying a large bag or camera case.
- Continual or loud arguing.
- Window cleaners: close the curtains or blinds.

Do not be afraid to explain these relevant points to visitors.

One reason why I believe that parrots and children are not a good combination is that few children are able to approach birds slowly with smooth movements. Some children might learn to do this but those who are not used to the company of parrots frequently cause a fearful reaction. If the children are members of the family, and permanently present, the parrot might be stressed.

As Kashmir Csaky so rightly says: "Signs of stress can be subtle and frequently go unrecognised... The most common stress indicators are nail chewing, leg biting, and sudden shoulder and chest preening. These displacement behaviours are normally done with an eye fixed on whomever or whatever is making the bird feel stressed. Displacement behaviours are either self-directed or directed at inanimate objects and occur when a bird is conflicted between two possible choices."

Aviary birds

For parrots in aviaries, females can suffer a lot of stress during the breeding season. Overall mortality among collections at this time can be unacceptably high. (Just look at all the "Wanted" advertisements for female parrots.) Not all keepers are astute or experienced enough to understand what is happening.

A friend e-mailed me about a six-year-old pair of Blue-headed Pionus Parrots *(Pionus menstruus).* The female was spending much time in the nest-box but not allowing the male to enter. When she left the nest the male would follow her. In the previous year the same thing happened and the female laid infertile eggs. This seemed to be a clear case of the male coming into breeding condition before the female. She was trying to avoid him by staying inside the nest-box. This could have led to her health deteriorating, not only from lack of food, but from stress. Breeders should recognise such stressful periods for a female and remove the male for a couple of months or more, until it is more likely that the female is in a receptive condition. Some male parrots become aggressive when the female does not respond. This, of course, is even more frightening for the female and could result in her death.

Stress as a new concept

When I gave a presentation on feather destructive behaviour in parrots to a large audience I emphasised that stress can be a major cause of this problem. A press photographer who had been covering the event came up to me afterwards. He told me that he had found my information extremely interesting. He had no idea that parrots could suffer from stress. That set me thinking abut how little the average person considers what is happening in the mind of a bird or animal. Many of these “average” people will become companion bird owners at some time in their lives. If they do not understand the concept that birds suffer stress they are not capable of caring for them in a humane way.

11. AGGRESSION AND BITING

Aggression in the wild is most likely to occur in defence of a nest, especially where such sites are hard to find (see **18. Territoriality**). In a forest fragment at Tobuna, in Misiones, Argentina, two pairs of the Endangered Vinaceous Amazons *(Amazona vinacea)* gave chase in flight. A dramatic scene ensued. Two of the opposing parrots grabbed one another and dropped at least 20m (66ft). Later on, two fell to the forest floor locked together. Neither pair nested in the forest fragment that year (Cockle and Bodrati, 2011). Perhaps both pairs were desperate to find a nest site but no such site existed there.

Phil Gregory once watched two Black-capped Lories *(Lorius lory somu)* at Ok Menga, Papua New Guinea. He told me that they were fighting so intently that they fell from their perch, high in a tree, and tumbled to the ground, squawking loudly and still locked together. They recovered and flew off. Actual fights among wild parrots, when bites occur, are probably rare except when a nest is being defended.

In our homes

Many parrots go through a biting stage, usually at around one to two and a half years. This can be difficult to cope with. Much patience is necessary. If you are always kind and loving, the phase will pass.

However, biting is often caused by human actions. Biting is the main aggressive behaviour that some parrots display after threat warnings such as the gape (see **7. Understanding Behaviours**). In my book *The Loving Care of Pet Parrots* I wrote: "Solving the biting problem is dependent on one factor: understanding your bird. Parrots bite most out of lack of discipline, fear, jealousy, territorial defence, neglect, as a result of being teased and, in mature birds, during what should be their breeding season. It is vitally important that biting should be stopped from the outset; if it is not, a relationship based on love and trust is totally impossible. The owner will fear the bird and the bird will be confused and lack respect for his owner."

Biting is an instinctive and almost inevitable form of defence for a captive bird in a small space. Paul Walfield bought a Red-bellied Parrot *(Poicephalus rufiventris)* from a pet shop. He was told it was two months old (if true, it was too young to sell). When he got "Louie" home it fell several times on to the floor. When his wife picked it up, it bit her severely on the thumb. Every time the poor bird tried to fly it fell to the ground and every attempt to handle it resulted in biting and charging at fingers. It apparently took the couple some time to realise that its wings "were clipped right down to the wing web". (The appalling practice of wing-clipping young parrots should be illegal.)

Was it any wonder that the little parrot was so aggressive? If a parrot cannot fly away from a threat, it bites. In time his flight feathers grew. His owner wrote: "Louie is a true aviator. Now that he has accomplished his goal [to fly], he no longer seems to have a need to attack. No longer does he charge our extremities with beak open" (Walfield, 2003).

This was written as though there was no understanding of the fear and frustration the little parrot had felt because it was unable to fly, that is, to escape unwanted human attention. Its only defence was to bite. In this case the reason for the aggression was obvious. In other cases it is not – and in the context of companion parrot and human, a person's fear of such an attack can ruin the relationship and often results in parrots being rehomed.

Many parrots bite because the behaviour of the carer is inconsistent. The parrot cannot understand what the person wants it to do – and it bites out of frustration. Biting is often the result of human behaviour which, to the comparatively small parrot, appears threatening or aggressive. The human probably does not intend this: he or she just does not understand what evokes reactions of fear. (See 10. **Stress and Fear.**) Lack of rapport with animals, or inexperience with them at close quarters, is easily detected by most birds. Those who know parrots interpret the behaviour that warns a parrot is about to bite. It is vitally important to know the signs because being bitten can be the start of a deteriorating relationship between human and parrot.

Inexperienced parrot owners can somehow miss or fail to interpret warning signs. These include the most obvious – the gape. Other signs are walking along the perch in a swaggering, exaggerated manner with head lowered and plumage sleeked and – in Amazons and some other parrots, dilating the pupil of the eye and flaring the tail. Leaning towards the person who is the perceived threat and lunging at him or her may be the last action before biting.

Invasion of territory or the likelihood of it, often triggers biting. My Yellow-fronted Amazon *(Amazona o. ochrocephala)* "Lito" was the most perfect companion I ever had. She maintained her sweet temperament during the 39½ years we spent together. But there was one thing she would not tolerate and which would have caused her to lunge at me and bite: invasion of her cage. My hand entering her cage, swiftly placed above her feet so that she stepped on, was acceptable. But woe betide me if I tried to do anything to the cage interior when she was inside it! I soon learned such an action was literally out of bounds.

Some parrots will nip when a hand reaches into the cage to take them out although in other circumstances nipping is not a problem. If it is necessary to move a parrot at a bad time, use a short perch, not your hand. Training a young parrot to step on to a perch as well as your hand, will pay dividends.

With Amazons, biting is often about a male in breeding condition assertively defending his mate or territory. Sometimes the "mate" is a human – and close attention from a spouse or another human triggers an attack. By this I mean more than a bite – flying at the face or attacking the feet. (Bare feet are especially irresistible!) Even chasing the unfortunate person. There are spouses who have suffered this on multiple occasions. This is foolish. If the parrot has the opportunity to attack, it is likely to take it. The bird's behaviour is not going to change. The only sensible solution is for the spouse to avoid such situations: not to be in the same room when the parrot is outside its cage. This is not "defeat" – it is common sense. Even a small parrot, such as a caique, can cause very nasty injuries.

Well-known American breeder Eb Cravens wrote: "It seems to me that belligerent attitude in captive parrots results from lack of respect. For some unfathomable reason, the closer to humans a hookbill becomes during the handfeeding and imprinting process, the more aggressive it may turn out to be, once it has broken its 'baby bond' and become emotionally independent" (Cravens, 2003). Most hand-fed parrots develop with no fear of people whereas parent-reared birds usually maintain a degree of wariness.

Infrequent and unexpected attacks are difficult to avoid but might be prevented by surreptitiously watching the parrot's behaviour – in other words, always being aware of its mood. There are times when a parrot has no wish to be handled. This might be in the early afternoon or the late evening which are the natural times for a parrot to be resting. Disturb it then at your peril! A bite is the likely result.

Other situations which can result in a bite are when the parrot is on your shoulder but not receiving your full attention. This is a dangerous place from your point of view – precisely because you don't have a point of view! You cannot see what it is doing or its reaction to a person nearby. If you value your ear-lobes, don't do it! You might be talking on the telephone or making the worst mistake of all – speaking to another parrot or to a human who is standing close to you.

Some actions by the carer are so foolish that a bite is inevitable. One woman was chewing gum and blowing gum bubbles when her Goffin's Cockatoo *(Cacatua goffiniana)* flew at her and inflicted a

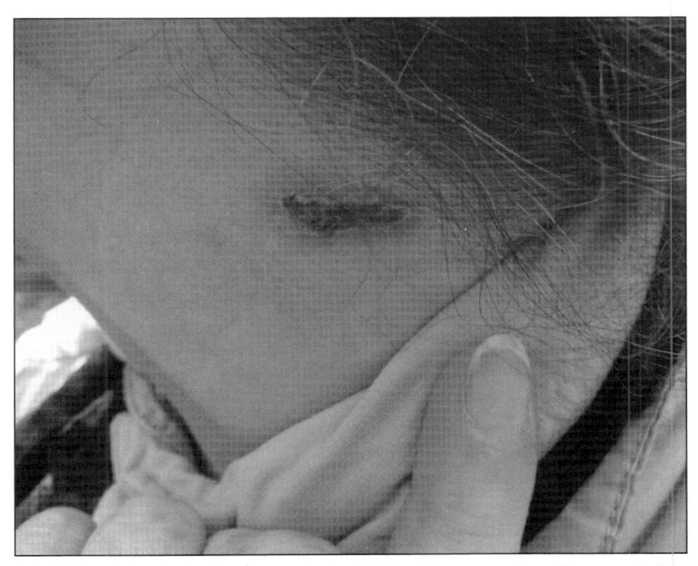

The neck of a caring owner, one week after a bite from a Moluccan Cockatoo. The bird suffered a sudden fright. It was not malicious.

Photograph: Parrot Welfare Foundation

serious bite on her chin. Afterwards she admitted that the cockatoo had been watching the bubbles and she teased the bird by sucking in the gum when the cockatoo came near. Any form of teasing is totally unacceptable and denotes an insensitive attitude.

Not your fault

It is often stated that there is no such thing as an aggressive parrot, the blame for all incidents being laid on the handler. Half a century with parrots has taught me that this is not true. Just as there are aggressive people, who might have inherited this undesirable characteristic, so there are some naturally aggressive parrots. I have one now. It and its unrelated female were acquired from the same breeder on the same day, when they were five to six months old, thus their life experiences have been identical. Now they have to be kept apart, in adjoining cages. I do not trust the male with the female and I need to keep my wits about me when feeding this bird, to avoid being bitten. In such cases, swing feeders or feeding hatches where hands are not exposed, are recommended.

Breeders should not use aggressive birds for reproductive purposes as this is the most undesirable trait that might be (and often is) inherited.

In aviaries

Aviaries in which more than one pair is kept are often the scene of aggressive encounters and even fatal attacks (see **18. Territoriality**). This sometimes occurs in zoos where a number of parrots are placed in a large aviary to make an exhibit that looks attractive. But unlike a flock of the same species, in which each bird knows its "pecking order", there might be a number of parrots of different species and sizes, some of which were hand-reared and never properly socialised. Bullying and aggression is likely to occur, especially where there are no mated pairs to fend off aggressors.

In breeding aviaries certain species are more liable to sudden and – to be honest – vicious attacks. These are caiques, cockatoos and certain Amazons, notably the excitable Double Yellow-head *(Amazona oratrix)** and related forms, such as the Vulnerable Yellow-naped *(Amazona auropalliata)**, and the Blue-front. A caring friend has kept a pair of Black-headed Caiques for more than 20 years. They are now past breeding age. She told me that the male has, on several occasions over the years, flown at her face. She showed me a bite on her arm from a recent attack, where she had lifted her arm to protect her face. Her explanation was that every so often he wants to remind her that he is the boss. Another friend has exactly the same experience with the same species.

Aggression in breeding pairs is often the result of the cage being too small. This means that the nest-box is too close to the feeders and the occupants feel threatened and protect the nest when the food dishes are changed. I would suggest that for medium-sized and large parrots there should be a minimum of 2m (6ft) between these two areas. Likewise, one reason why male parrots might attack their young when they leave the nest is because the breeding cage or aviary is not large enough – often *much* too small.

Aggressive behaviour of one member of a pair to another is not rare. These days CCTV cameras can be used to observe parrots when they are not aware of being watched. But their use cannot prevent fatal attacks, only warn that the behaviour of the male is threatening the female's life. However, aggressive behaviour by the male need not always put an end to a pairing. It can happen *after* the moult when only the male comes into breeding condition or before they would be expected to nest at the start of the breeding season. Removing the male from the aviary, until it is judged safe to return him (in weeks or months) is then essential.

* Not always recognised as separate species but recognition at species level seems advisable in view of their threatened status.

12. MEMORY

Several years ago I looked after a wild-caught Blue-fronted Amazon for several months. Although he had been in captivity for 30 years or more, I knew from his behaviour that he was not captive-bred. I could feel his sadness at being isolated from his own species. I showed him films of parrots in the wild. He saw Amazon parrots in Mexico, with shots of chicks being fed in the nest. Jasper, as he was called, became very excited and agitated and I guessed that he too had once reared chicks, somewhere in a hole in a tree in Brazil (he was of the nominate race), and that he had suddenly been captured, perhaps at a roost site, and torn away from everything he knew. The capture of wild-caught adult parrots is, to my mind, an unforgivable cruelty. How much do they remember of the life when they were free?

Parrots have extremely good memories, they can recall individuals, parrots and humans, after years of absence. This does not surprise me – I would only be surprised if they did not remember. From their reactions I know that individual parrots that I had not seen for years recognised me. Some reacted with such excitement that I was filled with pleasure. But many people express disbelief, almost, that parrots are capable of remembering and recognising humans after a long absence.

Research in the wild indicates that parrots "know" many other flock members. It therefore seems likely that they remember humans just as they do other parrots and that they can recall their days of freedom. There are many stories of parrots remembering people they disliked – and making this quite clear, even although they have not seen them for years.

I believe that most parrots have long memories. Why would wild parrots need this function? Many species are nomadic or they cover a wide area in a particular location in search of fruiting or flowering trees, great numbers of which are seasonal. It is of long-term benefit to them if they can recall the locations.

A clay lick much visited by tourists on the Rio Manu in Peru is used by nine species of macaws and parrots. Researchers observed them in order to judge their predator responses to various threats. The strongest responses were to certain raptors, jaguars and the Machiguenga people who catch parrots, or used to do so, as evidenced by macaw and parrot feathers in nearby villages. The parrots responded in a different way to researchers and to tourists.

The tourist industry there is mainly driven by the presence of the clay lick, probably the most famous in South America. At the time the research was carried out trading in parrots was diminishing there yet apparently parrots still perceived the Machiguenga as a threat (Burger and Gochfeld, 2003). This means that either long-lived parrots remember flock members being trapped by these people or, perhaps equally likely, the predator response to them has been passed from one generation to the next.

A friend told me a story about her Senegal Parrot *(Poicephalus senegalus)* that I found very interesting. Julia had acquired him straight from the house of his previous owner who had died. The Senegal was very thin and hungry so possibly had been without food for several days. One day, several years later, she felt very ill due to a migraine. It was morning and she went to her bedroom to rest. After a while she could hear the parrot behaving in a very agitated manner. She went to see what was wrong. When she left the room he again became very upset and did not calm down until she rested on the sofa where he could see her. Was he remembering the time when his previous owner was ill, went upstairs and never came down again?

PART III.
BEHAVIOUR

The best things in life

are birds...

and watching their

behaviour.

13. SENTIENT BEINGS

The dictionary defines sentient as "capable of feeling and perception" and "capable of responding emotionally". I believe that parrots have a complex range of emotions, including grief, contentment, love, fear, frustration (sexual and otherwise), depression, dislike and jealousy. The latter is often observed in companion parrots but is less likely to be observed in aviary birds –mainly due to lack of intense observation.

In parrots, much or all of the task of feeding the young after they leave the nest falls to the male parent. I recall an interesting case with a pair of Palm Cockatoos. They fledged their first youngster on November 18. The young one started to feed himself one month later but the male continued to feed him occasionally until February. I observed increasing animosity between the female parent and her offspring. On February 15 I saw her lunge at him and send him flying. The next day the male was again feeding him and, unusually, the female had placed herself on the other side of her mate and approached him as though she wanted to be fed. She was clearly jealous of the attention her son was receiving. He was removed to another aviary.

Tame birds will express anger, annoyance and impatience with a nip or a bite. Three days before I wrote this I returned from three nights away from home. As I let two of my Crimson-bellied Conures out into their flight next morning, the male landed on my shoulder and nipped my ear four times. This is a behaviour he has not shown since he was less than eighteen months old. He seemed to be expressing his displeasure at my absence. The food –including a wide range of foraging items –is not as good when I am away! (These birds are highly food-orientated.)

Are parrots considerate and caring?

Human beings often interpret parrot behaviour to mean that their bird really cares about them. It might be true or it might be wishful thinking. To decide whether parrots are sentient, caring beings perhaps we should study how parrots behave towards each other. Are they thoughtful, considerate? In captive birds we can study individual personalities which differ as much as do those of humans. Some parrots are gentle and caring; others are selfish and bullying.

An extreme example of a caring mate relates to a Sun Conure *(Aratinga solstitialis).* In South Africa Elaine Whitwain kept a prolific breeding pair. One day, when there were three young in the nest-box, the male fell off the perch, paralysed. The female then fed him as devotedly as though he were one of her chicks. Elaine put him back in the nest-box every night but every day he had to be caught up for an antibiotic injection. For the first few days the female attacked Elaine when she caught him. This was a long treatment and she apparently accepted the intrusion after a while. For an amazing six months the female fed her mate. Gradually he recovered to the degree that he was able to sire one more chick. The prolonged and devoted attention of the female can surely only be interpreted as intense caring.

The female Paradise Parakeet tries to arouse her dead mate.

The saddest story in this context made me cry when I first read it. Carl Lumholtz was a Norwegian explorer who spent four years in Queensland, Australia, in the 1880s. In that era most adventurers carried guns and enjoyed the destructive pastime of killing for "fun". Lumholtz encountered a pair of Paradise Parakeets *(Psephotus pulcherrimus).* These beautiful birds were walking near an ant-hill (possibly their nest site), eating grass seeds. He shot the male. He did not bother to pick it up at once but he did note the fine scarlet feathers of the abdomen, shining in the rays of the setting sun.

He saw the female fly down to her dead mate. With her beak she repeatedly lifted up his head, then walked to and fro over the body. She flew away but returned immediately with some grass stems in her beak and laid them before the dead bird, as if trying to persuade him to eat. She began again to raise her mate's head, and finally flew off as darkness was coming. Lumholtz wrote: "I approached the tree, and a shot put an end to the faithful animal's sorrow" (Lumholtz, 1902).

The man was apparently totally without sentiment so it seems unlikely that he invented the story of the female's compassion –an emotion of which he was obviously totally lacking. The story is all the more shocking because Lumholtz was watching a species which became extinct in the 1920s.

Fortunately, much has changed regarding human attitudes since Lumholtz's dreadful act of vandalism. The second half of the 20th century brought an elevated human awareness of birds and animals as sentient beings. While this awareness is increasing all the time, there is probably still a majority of humans who have scant awareness of what animals feel and experience. When I asked Mandy Beekmans, whose art so brilliantly enhances this book, to illustrate this sad story, she wrote back:

"I agree with your suggestion that probably a majority of humans do not (yet) recognize that animals are capable of having complex emotions. I sometimes feel that people either seem to wear blinders (actively denying complex emotions in animals) or that they anthropomorphise behaviour. This is a sensitive subject for me, and I support raising awareness of complex emotions in non-human animals."

Countless instances have been related of parrots and other wild birds watching over the dead body of their partner. It is a poignant sight which I have witnessed myself. In Australia, John Courtney told me that he had often seen Rosellas (*Platycercus* parakeets) sitting beside a dead mate on the highway near to his home.

On the Indonesian island of Flores, a Lesser Sulphur-crested Cockatoo *(Cacatua sulphurea parvula)* was shot from a flock that was raiding a corn crop. Its body was hung up, in a misguided attempt to warn off other cockatoos. The dead bird's mate returned to sit in silence close to the body (Schmutz, 1977). The pair bond between two compatible cockatoos is probably one of the strongest in the avian world. Cockatoos display affection towards their mates in a way which is very easy for the human observer to understand. The surviving cockatoo would have experienced the overwhelming emotions of bewilderment, shock and grief, rendering it unable to utter a sound.

Cockatoos are the most emotional, expressive and sentient of all parrots. Those that have been hand-reared develop an attachment to a human that can be almost impossible for that person to live with. They demand constant affection and attention. Unlike other parrots, they do not respond to food rewards: they only want physical contact. Cockatoos are unlike other parrots in that mating can occur at any time, not only as a prelude to reproduction. They can only be described as highly sexually aware and sensuous creatures –unlike nearly all other birds.

Parrots closely bonded to a human definitely know when that person is ill or sad. They grieve at the loss of their human companion or of one of the same species and can become very quiet and depressed for days or even weeks. They might cease to talk, lose their appetite and show the same symptoms of grieving as do humans. Many parrot owners who keep several birds in their home have noted how, when one died, the others became quiet and unresponsive.

In the USA a Yellow-naped Amazon lived with Jack and his wife Donna. When Jack was terminally ill with cancer the Amazon became "much clingier" with Jack and more demanding of attention from both of them. Perhaps he sensed that Jack was dying. On the day he died there were ten family members in the house. The Amazon did not make a sound, although normally he would be clamouring to be let out when there was company. During the following weeks when he was let out, he sat dejectedly on Jack's chair and Donna knew he was grieving along with her.

Also in Brazil, Kilma Manso's best friend and companion was a Blue-fronted Amazon. Kilma was deeply involved in parrot conservation and rehabilitation. After working for the police to apprehend wildlife smugglers, she helped to rehabilitate some of the thousands of parrots, mainly Amazons, that are taken illegally from nests every year. Kilma had a long and deep bond with her Amazon who was jealous of any other bird or animal which occupied her attention.

One day Kilma was feeding a confiscated Orange-winged Amazon *(Amazona amazonica)* chick that needed extra care. Her Amazon Porcina was at first jealous but the young parrot, illegally removed from the nest at a very early age, was not afraid of the aggressive behaviour. Instinctively he put his beak inside hers and pumped his head in the hope of being fed. To Kilma's amazement Porcina opened her beak and the young one immediately inserted his beak, soliciting food, and Porcina fed him. Kilma wrote to me: "After this she kept feeding him until he was able to eat alone. It was a fantastic experience for me –a great opportunity to see how parrots can care about each other."

Some people might say that the older Amazon was not feeding the young one for altruistic reasons but that the action was merely instinctive. Perhaps they would be right. However, as someone who had a very close bond with an Amazon for 39 years, I know that in the same circumstances my Amazon would have bitten a chick out of jealousy. Also, an Amazon that has lived with people for many years might not have a strong instinct to feed a chick.

In the UK, one caring person who rehomes parrots suffered the deaths of several conures and a parrot after one had almost certainly been in contact with diseased birds in its previous location. She told me that as one conure was dying its "best friend" held his wing protectively over it. When another conure became ill, the others in the group "formed a ring around her to protect her" when her owner had to remove her.

Another sad story relates to a pair of Rainbow Lorikeets *(Trichoglossus h.haematodus)* belonging to a friend. The male, her former pet, could speak a few words but seldom did so. The female, of unknown age and origin, was found dead but still standing on the edge of the swing feeder, with her faithful mate perched beside her. When her body was removed the male repeatedly uttered the words "Bye bye, bye bye" until he could no longer see her.

Perceptive and self-aware?

How perceptive are parrots regarding death or impending death? In the UK Donald Risdon was a respected aviculturist who founded The Tropical Bird Gardens in Somerset, a much-loved location at Rode, sadly no longer in existence. When the gardens opened, very few people in the UK had seen free-flying macaws. They were the stars. One of them was a Scarlet Macaw called Edwina. Betty Risdon wrote about her in her book *The Road to Rode:*

"She would always come over to our bungalow and knock on the French windows if either of us had been away or ill. One day she did her usual knocking and Don and I couldn't understand her message. We had not been away or ill. Edwina died that night. It was her way of saying goodbye."

This is not the only case I have heard of a parrot who knew it was about to die and was perhaps trying to communicate this –but most humans are not "tuned in" enough to other intelligent forms of life to understand. Some dog owners give true accounts of a dog that behaved in an unusual way to say goodbye just before it died, when there was no sign to a human that its death was near. This surely means that dogs, macaws and other parrots have a degree of self-awareness with which few people credit them.

Look into the eyes of a large macaw and you will find a thoughtful, sentient being.

In *A Century of Parrots* (2006) I wrote: "I have often looked into the eyes of a large macaw and felt troubled and guilty because I could feel that here was a creature as sentient as I am –and what right have we humans to shut them away from their own kind and the tender affection they give to each other, to stop them flying their graceful flight, to deprive them of their birthright: the rich environment of the green forests, the tropical warmth and humidity?"

These might seem strange words from someone with almost a lifetime's involvement with captive parrots, yet the passing years and numerous journeys to observe parrots in the wild have increased, not diminished, these feelings. Much as I love my own birds –and I cannot imagine life without them –my greatest joy is to see parrots where they truly belong, where they are free to fly wherever they wish.

We can never see into the minds of birds: we can only try to interpret what we see. But sometimes it is difficult to interpret an action other than it arises from one bird caring for another.

When a parrot transfers that caring to a human, and the love is returned in full measure, the result can be an enduring love that lasts for decades. Tiko is an Amazon parrot, described as a Red-lored

(*Amazona autumnalis* –sub-species *salvini*). He lives with Joanna Burger, a Professor of Biology at Rutgers University, and author of more than 14 books on bird behaviour. Tiko struck lucky when he was rescued by Joanna. He was thirty years old, his plumage was almost colourless, he received only parrot mixture, he was ignored and never let out of his small cage.

Joanna's ability to interpret the Amazon's feelings and behaviour, as related in *The Parrot who Owns Me,* makes this a book that every parrot keeper should read. It is an inspirational story of how she won the heart of a sad, neglected parrot. Her insight into what he needed will give entry to many people into a hitherto unknown world.

She wrote: *Our training drills into us [scientists] an aversion to anthropomorphic judgments. I once considered it the epitome of bad science to attribute human thought, feelings, and language ability to animals. But over the years I have changed my mind. I have come to regard it as at least equally benighted to automatically assume that animals lack these qualities.*

Tiko was 46 years old when she wrote about his life. She observed that the capacity of animals for intimacy and connection with one another and with us is not fixed but grows and develops. It can, perhaps, be taught. It is influenced by experiences and events. This is another facet of animal intelligence which seems to apply especially to parrots.

Joanna's relationship with Tiko grew and developed over the years. In July 2014 she told me: "Tiko has just passed his 60th birthday. He has arthritis and a cataract in one eye. I love him dearly and no matter what else I am doing, he is with me, placing his head against my leg, foot or hand, and calling softly. He has learned to adapt to his arthritis, and readily climbs the ladders I give him to reach his favourite perches, and eats the fruit with his medicine on. I take him out daily for about fifteen minutes so he can get a little Vitamin D. He is amazing, happy and very loving. And like all of us, he is adapting to getting old, and making the best of it, while keeping his sense of humour."

In *The Human Nature of Birds* Theodore Xenophon Barber wrote: "Since avian intelligence and awareness is a factual, demonstrable conception, it cannot be squashed and is bound to prevail... The forthcoming revolution in human thought will be led by men and women who are no longer intimidated by the taboo against perceiving birds as conscious individuals... and who transmit to others (via lectures, demonstrations, television shows, news articles, and group projects) their perception of the humanlike characteristics of birds. The avian revolution will be complete when the new generation accepts as natural that people and birds can understand each other and relate to each other not only as equals but also as friends."

Written in 1993, Barber's words about the revolution in human thought are only slowly resonating –on a planet where most people are either struggling to survive or are affluent but have little interest in the natural world. Whatever their status, by striving to understand more about the lives and feelings of the incredibly wonderful feathered creatures that share our planet, their own lives would be enriched beyond measure.

14. INTELLIGENCE

What is intelligence? It includes problem-solving, learning new skills and emotional intelligence. And even tool-making.

The cerebral cortex is the part of the brain which is considered to be the main area of intelligence in animals. In birds this area is relatively small; however, birds use a different part of the brain, the medio-rostral HVC*, as the seat of their intelligence. At the University of California, San Diego, neuroscientist Harvey Karten discovered that the lower part of the avian brain is functionally similar to that in humans.

Why are some groups of birds more intelligent than others, and why do some parrot species show a greater degree of intelligence? Scientists believe there are probably two determining factors: the foraging behaviour (how cognitively demanding it is to obtain food) and the social complexity of that species. If this is true, it helps to explain why the most intelligent species are the most susceptible to feather plucking unless they have foraging enrichment.

Research

At Hamburg University Dr Ralf Wanker studied Spectacled Parrotlets *(Forpus conspicillatus)* and other parrots. He stated that greater complexity in the social system of a species favoured the evolution of mental skills in working out logical connections and in solving problems. Researchers tested four parrot species with different social systems and diets: Spectacled Parrotlets, Green-winged Macaws, Sulphur-crested Cockatoos *(Cacatua galerita)* and Rainbow Lorikeets.

The scientists believed that one of the characteristics of complex cognition is the ability to understand cause and effect. One way of testing this was to offer the birds an out-of-reach reward. Five variations on a string-pulling task could lead to a food reward, such as a pair of crossed strings. Only one delivered food but pulling the string directly above the food would not access it. The Spectacled Parrotlets and the Rainbow Lorikeets worked this out but the macaws and cockatoos did not. Only the parrotlets found out how to get the food when the strings were the same colour. Most people who know all four species would, I think, be surprised by this result, as they consider cockatoos and macaws to be far more intelligent than sparrow-sized parrotlets.

The next test was designed to probe flexibility of behaviour. The various parrots were offered longer strings so that they could obtain the food from the ground rather than by pulling up the string. Several members of all four species adapted their problem-solving strategies by getting the food from the ground but only the parrotlets and lorikeets preferred this method. In my opinion, this is not surprising as most macaws are canopy-feeders, not ground feeders, so they instinctively avoid going to the ground. Although the lorikeets feed mainly off the ground, they do descend to

*HVC = high vocal centre: a nucleus in the brain of songbirds (Passeriformes) necessary for the learning and production of song. Parrots seem to have structures similar to HVC.

drink and sometimes to forage. Perhaps the ability to solve problems was not tested in this case but natural behaviour was more significant.

Another task involved a string with a reward on the end and another string with a reward under it but not attached. The parrotlets were the only species to solve this task.

When the results of the tests were compared with the birds' lifestyles, they found the differences were best explained by the species' social structures rather than their diets. The parrotlets live in what is called a fission-fusion society. In their large groups they form different social sub-units that split and merge, providing the opportunity for many different kinds of social interactions. The researchers said that the parrotlets were the only one of the four species tested that forms crèches where young birds pass through the socialisation process. The macaws and cockatoos live in small, stable family groups centred around a breeding pair and their offspring. The social organisation of Rainbow Lorikeets falls somewhere between the parrotlets, and the macaws and cockatoos.

It was believed that the parrotlets' cognitive performance was best because individuals have to recognise group members. (But surely this applies to most species.) "These demands favour them understanding functional relationships, such as which actions cause which outcomes" (Krasheninnikova *et al,* 2013).

Opinion
It is an interesting theory but perhaps one that needs further research. It would have been revealing if the Kea *(Nestor notabilis),* the charismatic mountain parrot from New Zealand, had been included in the test group. It probably would have blown this theory out of the water! It lives in small social or family groups, and is generally considered to be the most intelligent and mischievous parrot on the planet. It enjoys damaging vehicles for recreational purposes! It can solve logical puzzles, such as pushing and pulling things in a certain order to get to food and, more importantly, it can work with other group members to achieve an objective.

The Kea's beak is used for mischief.

Social life has been considered to be an important factor for the evolution of intelligence. In this case the Kea's very harsh alpine environment must be the driving force. It must use its initiative to survive, exploring every nook and cranny to find food, using its long beak. It also uses its strength. One Kea was filmed removing two large logs – as large as itself and probably a good deal heavier – from the top of a wheelie bin, then opening the bin to find discarded items of food.

It is often stated that parrots are among the most intelligent of birds. But how do we define intelligence? Some scientists include play and co-operative breeding among the factors that denote intelligence. One interpretation (Perrin, 2012) is that intelligence is flexibility in transferring skills acquired in one domain to another. Thus, states Prof. Mike Perrin: "...long-lived parrots existing in complex social systems, not unlike those of some primates, use abilities honed for social gains to direct other forms of information processing. Parrots are able to distinguish non-predators from predators or poisonous from healthful foods, to recognise and to remember environmental regularities and to adapt to unpredictable environmental changes over an extensive lifetime."

While this is true, it could equally apply to hundreds of other species, including the Great Tit *(Parus major).* The main difference is that tits generally have a short life. Probably no one would cite this tit as an example of a highly intelligent bird, yet it is. (See page 78).

My point is that because countless parrots live in close association with man, their cognitive abilities are well known. I believe that many birds are very intelligent – they need to be to survive in a world full of hazards, many of them man-made. Parrots and crows are among the most highly intelligent of birds and, among parrots, I would suggest that Keas, Greys, cockatoos and macaws are exceptionally intelligent.

Perrin suggests that the combination of intelligence and advanced communication skills has arisen in parrots and directs not only learning but also what is appropriate to learn.

I would suggest that the ability to solve problems is a good indicator of the level of intelligence. White cockatoos (*Cacatua* species) excel. In Australia the Red-capped Parrot, also called Pileated Parakeet has a lengthened upper mandible (see illustration on page 3). This enables it to eat the large seeds of the eucalypt marri *(Corymbia calophylla),* locally known as honkey nuts. These lie within a hard, woody capsule that measures up to 5cm (2in) in length. It is widespread and common in south-western Australia.

Both Long-billed Corella and Little Corellas have been introduced to this region. These corellas are not known to feed on eucalpyts, except on the seeds in the very small capsules of the river red gum *(Eucalyptus camaldulensis)* in south-eastern Australia, where this tree and the cockatoos are common. Both introduced species have devised a technique to tip up the head and empty the seeds into the bill when they find opened marri capsules (too hard for even these cockatoos to open) which are at the appropriate stage to release the seeds. This cannot be instinctive behaviour because marri does not exist in their natural habitat. The cockatoos occur in mixed flocks in Western Australia so it seems likely that one species learned the technique from the other (Burbridge, 2008). However, they had to learn a new skill to be able to utilise a food source. This surely is intelligence.

In the same vein, in 2012 researchers near Vienna discovered that a captive-bred Goffin's Cockatoo called Figaro also learned a new skill to acquire food. Amazingly he achieved this by tool-making. One day Figaro was playing with a pebble when it dropped outside the welded mesh of the aviary. After unsuccessfully trying to reach the pebble with his foot, he used a stick on the aviary floor. The researchers (from Oxford University and the Max-Planck Institute for Ornithology in Germany) then tested Figaro by placing nuts just outside his enclosure on the timber along the front of the aviary. He made his own tool by splintering off pieces of wood from it and then trimmed them to the right size and shape to rake in the nuts from outside. This exercise was repeated in ten trials over three days. The time it took him to manufacture suitable tools decreased with each success (Auersperg *et al*, 2012).

Tool-using has often been cited as an example of extreme intelligence among animals. Indeed, at one time it was thought that only man and chimpanzees could do this. It is now known that a number of mammals and birds use tools, including, famously, the New Caledonian Crow *(Corvus moneduloides)* and the Woodpecker-finch *(Camarhynchus pallidus)* of the Galapagos. I suspect that quite a few parrot carers could cite instances of birds using objects as tools.

In the USA Dr Stuart Metz has studied the Moluccan Cockatoo, wild and captive, for many years. He told me: "I have learned to never underestimate what cockatoos can pull off. I have seen some acts of intelligence and motor skills which have been amazing. But this one might just 'take the cake'. Wearing sneaker-type trail shoes, I was sitting in a chair playing with Ipo, my male Salmon-crested cockatoo. Ipo was on my knee. Then, out of the blue, he preceded to untie the knot on one shoe. What was most amazing is that he did that with almost no hesitation at any point – and that it had been a double knot. (Some might call it a triple knot – which not infrequently I have my own problem figuring out!) Both the finesse in his fine motor skills, as well as his deciphering of a pretty complex "puzzle" left me amazed, to say the least. Then I was sitting, probably with my jaw dropped, when to my amazement, he did the exact same thing on the other shoelace. You'd have to see this to believe it, and even then, you might not believe your eyes!"

This trick involved acute observation and dexterous manipulation of beak and tongue – and was not learned from instruction. Many parrots can be taught to do various tricks and most learn quickly – usually with food rewards. In India, people used to teach Ringneck Parakeets to perform for money in front of groups of people. One such bird, bought in Calcutta in 1910, could even twirl a stick which had been lit at both ends – demonstrating prowess as well as lack of fear of fire. He could also use a little bow and arrow, draw water in a small wooden bucket, thread tiny beads and fire a miniature cannon which went off with a loud bang. Not everyone agrees with parrots entertaining the public but the fact is most avian performers enjoy the attention and the variety it adds to their day.

In *The Wisdom of Birds* Tim Birkhead states that two groups of birds stand out as being particularly clever: crows and parrots. He writes: "What is special about crows and parrots? The answer is that they live in complex environments where behavioural flexibility is the key to survival. In facts, in terms of their behaviour crows and parrots are more like primates – monkeys and apes – than birds."

Birkhead states that corvids (crows) and parrots have relatively large forebrains and more sophisticated cognitive skills and behavioural flexibility than any other groups of birds" (Birkhead, 2008).

When David Attenborough made a list of "the 15 smartest animals on earth" he ranked them for their ability to use creative thinking. The top bird was the crow at number five, followed by the African Grey Parrot at Number 6. This was his opinion – not backed up by scientific evidence!

In *A Century of Parrots* I wrote: *During the last decade of the 20th century the intelligence of parrots and their capacity to interpret facts and even to understand abstract thought was brought to the attention of many people worldwide who had previously attributed them with a much lower level of intelligence.*

The work of Irene Pepperberg in the United States, with the Grey Parrot Alex, was responsible to a great degree as he received so much coverage in the Press internationally for his accomplishments. Alex was believed to be extraordinarily talented. However, probably the results with any one of thousands of other Grey Parrots would have brought similar findings. It was the teachers and their intense methods that influenced Alex's learning capabilities. No other parrot in history has received similar tuition. Alex could count up to six, identify and name (speak the words) colours and objects and perform many other tasks.

Alex was a laboratory bird from 1973 until in September 2007 he was found dead in his cage. Cause unknown. He did not live in a loving home. He plucked himself and he died prematurely. Yes, he received a lot of attention in the form of eight to 12 hours daily tuition. But did he have a good life? Apparently he was able to express frustration at repetitive scientific trials. Imagine if a child, grown to an adult, had this level of one-to-one tuition, day in, day out, for thirty years. I feel that it, too, would be tearing its hair out.

Birdbrain?

My dictionary defines "birdbrain" as someone who is silly or mildly unintelligent. The term indicates that humans consider birds to be lacking in intelligence; in my view it indicates ignorance on the part of those who use this expression. When you consider that a Grey Parrot weighs much less than half the weight of a human brain (1.3 to 1.4 kilos), the intellect and cleverness of parrots and other birds is extraordinary. It is time more people gave them credit for the remarkable creatures they are.

15. COMMUNICATION: VOCAL AND OTHERWISE

Parrots communicate intent and information to each other, through vocalisations and body language – just as we humans do. One of the most common vocalisations is the contact call. This is given when the flock or one of a pair is about to take flight, when individuals become separated from the flock and when two birds are reunited after a separation. Persistent contact calls reinforce the cohesion of a flock and act as protection against predators.

Vocal communication

Parrots are not renowned for the mellifluous notes! However, a surprising number of species have pleasant and even melodious voices. Surely nothing can surpass the lovely loud thrush-like song of the Purple-bellied Parrot *(Triclaria malachitacea)*? Apart from a harsh *chack chack* given in moments of excitement, every note in its repertoire is musical. Both male and female have a soft sub-song and the female also sings, but with less intensity and less often.

The Lesser Vasa Parrot has a pleasant whistling song and the German name for the Red-rumped Parrot *(Psephotus haematonotis)* is *Singsittich.*

One study of Quaker (Monk) Parakeets identified eleven types of calls. Scientists suggest that most parrot species have ten to fifteen distinct vocalisations, each having a particular meaning. However, for many species, this might be an underestimate. Researchers documented nineteen different calls for the Eastern Rosella *(Platycercus eximus).*

Grey Parrots – acoustics

Few studies have been made of acoustics or vocal repertoire of wild parrots. In 2004 Diana May made recordings of Grey Parrots in the Central African Republic and in Cameroon. These were analysed using a spectrograph at the University of Arizona. It was found that their calls could be subdivided into thirty-nine acoustical types in the categories tonal, harmonic, noisy-harmonic and noisy. The majority of the calls were of the pure tonal call.

Alarm calls

One of the most obvious vocalisations is the alarm call which is usually easily recognised by its urgency and persistence. Parrots can also respond to the alarm calls of other birds, and mammals such as monkeys.

One incident in my aviaries was interesting. It happened early one morning when I went to let into the outside flight my family group of Crimson-bellied Conures. A Blackbird was perched on the television aerial on the roof of my house emitting its alarm call, on and on and on. The senior male conure and his mate came to the exit and perched on the edge. Instead of the usual family eruption (it is like pouring milk from a jug!) the senior pair looked around warily, craning their heads upwards, and prevented the exit of the rest of the family. When they were satisfied that there was no danger in the vicinity, they flew into the flight, followed warily a few seconds later by the two one-year-olds, then by the four recently fledged young.

Identifying family members

Denis Saunders studied the large black Endangered Carnaby's Cockatoo during a period of 27 years – a highly significant and almost unique study (Saunders, 1983). This magnificent bird, found only in Western Australia, is now very seriously threatened by loss of habitat. Saunders identified fifteen types of vocalisations. He was probably the first person to discover that a female parrot, when incubating, responds only to the contact calls of her mate. She leaves the nest, flying to a nearby tree, on hearing the calls, knowing that the male has arrived to feed her. If she was unable to identify his calls she would leave the nest unnecessarily, either in response to another male or at random times, leaving eggs or chicks vulnerable to predation. Or perhaps she might miss his visits if he was reluctant to enter the nest hole.

Since this study, it has been found that all species on which research has been carried out, can identify the voice of their mate. I don't find this surprising. Humans can identify the voices of dozens of people and I fail to see why intelligent, alert creatures like parrots should not have the same ability. Indeed, it was found that Spectacled Parrotlets studied at Hamburg University in Germany use different calls like names to address specific family members.

Studying the tiny *Forpus* parrotlets to learn more about parrot vocalisations and behaviour started by accident. In July 1985, while studying Snail Kites *(Rostrhamus sociabilis)* on a cattle ranch in Venezuela, Steve Beissinger noticed a pair of Green-rumped Parrotlets *(F. passerinus)* nesting in a hollowed-out fence post. He decided to put up a nest-box to find out if they would use it. Now there are more than one hundred nest-boxes at this site, resulting in a wealth of information about parrotlet ecology and life history. Later the focus was on their communicative skills.

Many parrot species occur in flocks, thus vocal communication is necessary for individuals to stay in contact with the flock or with a mate.

In some species, such as *Poicephalus,* the calls of male and female are different, so the opposite sex can be identified in birds which are out of sight. The structures which produce sound are shorter in females (Perrin, 2012).

Conforming with group members

In recent years several research projects have shown how calls in individual birds can change when they are introduced to a new group. In other words, their calls conform to those of their companions. At the University of St Andrews in Scotland, a new individual with a different social call was introduced into each of three well-established trios of male Budgerigars with the same vocalisation. In all cases the call of the new bird changed to conform with that of the others without marked changes in the calls of the trios. The major change took place fifteen to thirty days after introduction.

Too noisy? Try these tips!

Calling occurs most often at dawn and dusk and is usually directed towards a close family member. Many captive parrots are especially noisy at these times and their calls are directed towards members of the human "family". It is unreasonable to expect a parrot never to have periods of loud vocalisations. Anyone not prepared for this, should not chose a parrot as a companion.

Parrots in the home may be noisier when they do not reside in the most-used room in the house. They call a lot as they would to a mate out of earshot. Note these points:

- A cage on castors is a wise choice so that it can be easily moved from one room to another. When your parrot knows you are close and can see you, he will not feel the need to call out so often.
- If your parrot is excessively noisy, the worst thing you can do is to shout at him to be quiet. He has then gained your attention – just the reaction he hoped for! He will probably scream even louder. The best advice is to leave the room without even glancing at him. Note: is there anything sadder to the ears of a bird lover than a parrot that yells: "Shut up!"
- To reduce excessive calling withhold a favourite item of food until the parrot is quiet. Hopefully he will come to associate being quiet with this "reward".
- If you can anticipate when your parrot will be noisy (such as when you leave the room) give him something to gnaw, such as a cardboard box or carton. Apple and willow branches are even better. Of course, such items should always be in the cage, but a new one is usually the most attractive. Or try a rolled up newspaper pushed through the bars or, for a large parrot, a nut that will not be quick to crack!
- Provide enrichment toys. Even simple ones at which the parrot must work to retrieve a food item are effective.
- If you have more than one parrot, ensure that each one has an equal share of your attention.
- If none of these suggestions are successful, get out the sprayer (plant mister), speak kindly to your parrot and spray it. This keeps it busy preening for a few minutes.

Parrots do not always scream because they want attention. Often they are responding to loud noises such as the vacuum cleaner, radio or television. You can quickly find out if this is the case.

Non-vocal communication

As yet, not many species have been studied in the field but results indicate that parrot communication is much more complex than was previously known. Scientists who observed Orange-winged Amazons in Pará, northern Brazil, were surprised to discover that males at the nest "gesture" to females. It was suggested that the sequence of gestures warned of possible threats towards offspring and are performed in silence to avoid alerting predators. Further observation would be needed to confirm this.

Sixteen nests were observed, covering the whole breeding season. Visual displays were recorded using digital cameras. The male apparently signalled to the female that it was safe to approach the nest by raising himself up and quickly bending his head back to about 45°, until the female approached and looked towards him. He then performed a downward head movement to encourage her to enter the nest. When she did not enter he moved his head down again, or moved to a more visible perch and repeated the bowing movement until she entered the nest.

The male used another gesture to encourage the female to leave the nest. He perched in the nest tree and moved his head down until the female left the cavity or until she put her head outside. However, in most observations of females exiting the nest the male emitted a specific vocalisation until she left (Moura *et al,* 2014).

One of my favourite aviaries, when I was curator at Loro Parque, Tenerife, was a large, high enclosure measuring about 12m x 6m (40ft x 20ft). It housed a group of Moluccan Cockatoos (wild-caught in those days). Watching the behaviour of the flock of fifteen was absolutely fascinating. On occasions, these highly intelligent birds would assemble on the ground in a group of six to eight, and stand together as though in discussion. I did not hear any vocalisation but they seemed to be communicating something...

Courtship and threat displays

Courtship display is, of course, a form of communication between male and female – the male informing the female that he is ready to breed, and vice versa. This display varies significantly in different groups of parrots. In Australian parakeets the male of some of the larger species will run up and down the perch near the female, in an agitated manner. In Rosellas and *Barnardius* species the male will stand more upright than usual, squaring his shoulders, spreading his tail and moving it from side to side. The male Eclectus Parrot *(Eclectus roratus)* opens his wings in display, to show the female the red underside.

In some species the display may be designed to show off areas of plumage that are fluorescent under UV light. These may contrast to adjoining feathered areas but the human eye cannot detect this (see **5. The Eyes have it!**).

Male Red-tailed Black Cockatoo, with crest erect and cheek feathers over the beak.

White cockatoos display with loud calls, wing-spreading, head-bobbing, crest raising and tail fanning. In the male Red-tailed Black Cockatoo the tail has very eye-catching wide bands of colour, as distinct from the barred and spotted tail feathers of the female. (See illustration on page 17). In display the male emphasises three areas of

The amazing threat display of the Hawk-headed Parrot.

plumage: the tail is spread, the crest is raised and brought forward on to the upper mandible and the cheek feathers are puffed out and brought forward to cover much of the bill.

This means that the bill, which is black in the male and white in the female, is almost hidden. It raises the question: Why is the beak colour different in male and female? It would be easy to say that inside the nest the white beak is more easily seen by a chick, but males also share in feeding the young.

Some parrot species use their plumage in threat display, the most notable example being the Hawk-headed Parrot *(Deroptyus accipitrinus)* whose bizarre headdress seems to be designed specifically to instil fear in other creatures. The feathers of the nape lie flat, looking very pretty, being red tipped with blue. When the parrot is angry or afraid, up go these feathers to form a crest quite unlike that of any other parrot – a circular fan that frames the head, making it look huge. But this is not all. The parrot's personality seems to change with equal rapidity. It becomes a hissing, lunging aggressor with blazing yellow eyes.

Macaws threaten by holding their wings high and open, thus doubling their body area and making them look even more massive than they are.

Communication between birds and humans

The casual human observer of birds, wild or captive, often feels puzzled. Why are they acting in a particular way? In comparison, intense and/or long-term studies of groups or individual birds, can reveal unexpected and sometimes truly amazing results that should make us all re-evaluate our attitudes towards them.

The classic works of Len Howard were way ahead of her time. In her magical books, *Birds as Individuals* (published in 1953) and *Living with Birds* (1956), Len Howard demonstrated spatial awareness and intelligence in a number of small species, truths about birds that were seldom recognised previously.

She was able to do this because she gained the trust of many birds in her garden.

Tits, especially Great Tits, were very dear to her. She studied numerous wild, free-living individuals, often over many years, and named each one. Twist was so called because she would roost with her tail pressed sideways, causing it to twist. She was unusual in that she seemed to understand some words. As a full-grown fledgling she would take naps perched on Len's knee or preen her feathers while perched on her hand. Len could raise her hand close to her face and gently rub down the tit's back. (Most birds dislike being touched on the back.)

One day she was astonished when Twist responded to 'Give me a kiss' by touching her nose with her beak. She thought this was chance – but next day she responded to the words in the same way and continued to do so for the rest of her life. If Len asked her a second time she would look at Len with a puzzled expression, not responding, and if she tried to coax her, she glanced at Len very crossly. After an hour or two she was again responsive. When Twist was hungry and Len had cheese for her, Twist would give three hurried kisses in succession, but even when tempted with cheese, she would not comply if asked twice running. She never offered a kiss without being asked and other words did not produce one.

Stacey O'Brien is an American Biologist who described her unique relationship with a Barn Owl in *Wesley, The Story of a Remarkable Owl.* She reared him, an abandoned owlet, and he shared her life and her home for 19 years. Why am I including this story? Because, despite the owl's status in mythology as a symbol of wisdom, most people would consider that a parrot was more intelligent and sentient than an owl. But captive birds are like children, how they are brought up and whether or not their carer is sympathetic and responsive to their needs, shapes their behaviour.

Stacey had always talked to Wesley, explaining things to him. She wrote: "An all-or-nothing creature, Wesley's emotional responses were as transparent as a child's. Things were black or white, good or bad, safe or dangerous, so he always expected me to follow through with exactly what I told him I was going to do." (Remember Stacey was a scientist, not a sentimental pet owner.) "Otherwise he'd become visibly upset, refuse to make eye contact, or even screech in protest until I did whatever I promised. Owls do not tolerate lies. If I said, 'I'll play with you in two hours, Wesley,' and then went about my business, two hours later he would start screeching and become unmanageable. I was stunned to learn that he had processed what *two hours* meant. I would say 'in two hours' and then follow up two hours later, and did that so often that he was eventually able to figure out that it meant he had to wait a certain period of time."

This was the only time-span he understood but he also learned what tonight and tomorrow meant. "Wesley would screech all through the night if I did not keep a commitment, making it impossible for me to sleep" (O'Brien, 2008).

What lessons can we draw from these happening in relation to parrots? Surely, if genuine communication between an owl and a human is

witnessed the killings. Reputedly, the dead woman's husband, a newspaper editor, read out the name of each suspect to the parrot. It was quiet – then squawked at the name of his nephew Ashutosh, shrieking and flapping its wings. Police said that the named man confessed.

The other story concerned a driver in Mexico who was stopped by the police during a routine alcohol check. Reputedly, his parakeet who was in the car, called out "He's drunk! He's drunk!" It was said to be true and the man was arrested – and released the next day. I do not believe this story because it seems unlikely that the parakeet would understand what "drunk" meant – even if he had heard it repeated several times.

I noticed, some years ago, how visitors to my house claimed, in my presence, that one of my birds uttered various words. But these words were just a figment of his or her imagination! People are very willing to credit parrots with statements they have not made!

Bizarrely, however, in Buenos Aires in 2006, judge Osvaldo Carlos ordered Pepo to be sent to prison. Two neighbours, disputing the parrot's ownership, were locked in a fierce custody battle. The judge directed that the parrot should remain under police "interrogation" until he spoke the owner's name. After five days Pepo obliged and then started to sing the anthem of the man's favourite football team! He was released!

Speech – not just mimicry

If we use the word "talk" to mean just that, rather than mimic, the Grey Parrot again comes out top. There are so many stories that prove these parrots, with their remarkable level of intelligence, use words correctly. They can, for example, refer to items they want or name objects for other reasons. A friend told me about one of her two Grey Parrots, which lives in the house with five or six dogs. One Grey calls them by name, never making a mistake and, as parrots outlive dogs, he has seen them come and go – and never again calls the name when one dies. One of the dogs is disliked by the parrot who ignores it, and never speaks its name. However, we should not conclude that other parrots cannot be equally clever. With so many Grey Parrots kept as close human companions, there are certain to be more anecdotes relating to them.

Many parrots use words in the correct context if their owners interact intelligently with them. As a quite ordinary example – and surely hundreds of owners could tell similar stories – one Grey watched his owner make a cup of tea. When the kettle whistled, the Grey said "Whaaat's that?" and was told "Cup of tea". He paused, then said: "Want a drink of water?" Thereafter, whenever the kettle boiled he said: "Want a drink? Have a cup of tea" and made loud slurping and gulping noises. In the evening when he was tired and wanted to return to his cage, he would say: "Sleepy…bird. Want to go night night."

The most interesting aspect of parrot mimicry is how some parrots can use a word or a phrase in one context and apply it correctly to another situation. At times it could be coincidence but there are many examples often very amusing ones – that show that they understand the word and are able to apply it appropriately out of its original context.

Companion parrots often mimic the telephone or the ping of the microwave oven. This is probably because they see a human hurrying to that object when it makes a sound. Perhaps they reason that if they copy that sound, the human will hurry to them! Or perhaps they just amuse themselves by learning and repeating a sound that appeals to them.

Learning the calls of their own species

In South Africa a study of Meyer's Parrots *(Poicephalus meyeri)* was carried out to determine whether contact with parents was important when young birds were developing their sub-songs (Masin *et al,* 2004). The latter are vocal trials in which birds practise songs. The researchers recorded vocalisations of pairs and their young – about 50 sub-song recordings for each young bird. Spectrograms were made to isolate and classify notes. Young showed 20% similarity to adult songs during the first week after fledging, suggesting that vocal learning begins very early. By the fifth week sub-songs matched 100% with their father's songs. Comparing the vocalisations of hand-reared young with the father's themes resulted in no common features for the whole weaning period. The calls of isolated chicks were mainly modified begging calls.

Are parrots hand-reared in isolation from their own species able to learn the correct vocalisations? This seems to be dependent on the individual, the species or the circumstances: there is no definitive answer. I hand-reared a Dusky Lory *(Pseudeos fuscata)* from the age of 22 days – with a Red-bellied Parrot. Both species are approximately the same size but totally unrelated. When weaned, the Dusky was kept with other lory species, but not its own, until it was eleven months old, when it went to a new home. It was with me all that time, during which the only sounds I ever heard it make were food-begging calls.

The vocalisations of the Stella's Lorikeet seem to be inherited, rather than learned. This species makes an unusual soft, drawn-out nasal sound which is unique to it and the closely-related Josephine's Lorikeet *(C. josefinae)*. A male Stella's hand-reared from the age of four weeks did not have visual or vocal contact with other members of its species until it was just over six months old. At 14 weeks it started trying to make this adult vocalisation which usually takes some months to perfect. It learned to make the sound at about the same rate as parent-reared young. Even at about eight months Stella's are not as adept as adults in this respect.

Some parrots with less complex calls, such as Amazons, appear to grow up making the normal vocalisations, even if hand-reared in isolation.

I know a hand-reared Lesser Sulphur-crested Cockatoo, kept near other cockatoos, that never shrieks. The only sound it ever makes is a soft *Wow*. This is surely the consequence of it being hand-reared in isolation from its own species. The other possibility is too horrible to contemplate. There is a terrible practice called devocalisation in the USA. I will not describe how this is done as I do not want readers to throw down this book in anger. In the leading avian veterinarian handbook it states: "The authors and editors consider devocalisation a cruel and unethical practice; therefore a procedure will not be described. Birds with vocalisation patterns that are unacceptable to a client should be placed in a new home" (Ritchie, Harrison and Harrison, 1994). British veterinary organisations considers this to be unethical and inhumane. I could find much

Dusky Lory and Red-bellied Parrot hand-reared together.

stronger words for any vet prepared to carry out such an operation…

A much-studied wild population of Green-rumped Parrotlets in Venezuela revealed something new to science. In 2007 and 2008 Karl Berg from Cornell University and his researchers set up inconspicuous video cameras and audio recorders inside and outside seventeen nest cavities made from PVC pipes, with welded mesh inserts for inspection purposes. They found that each adult had its own unique contact call which was more similar to its mate's than calls made by adults at other nests. The nestlings' voices were, unsurprisingly, notably similar to those of their parents or, in the case of one nest, foster parents, thus it seems that the young learn from their parents (or foster parents), not as the result of biological inheritance. It is likely that this vocal recognition restricts parental care to its own fledglings before families move to communal foraging and roosting sites.

Why do parrots mimic?
Wild parrots imitate the calls of others of their kind when they move to a different area and join a new flock. This has been demonstrated in Yellow-naped Amazons *(Amazona auropalliata),* for example, which in captivity are exceptionally good mimics. The advantage for wild birds is that they will be accepted into a new flock if their vocalisations are the same. Several roost sites can share a dialect. Some Amazons at roosts bordering two dialects can use both calls appropriately so that they will be accepted at both roosts. (See also **15. Communication:** Vocal and Otherwise.)

Musicality
The musicality of certain species of Amazon parrots is nothing short of extraordinary. The closely-related Yellow-naped, Double Yellow-headed and Yellow-fronted (Yellow-crowned) Amazons have an ability to enjoy and imitate music which, I believe, is not equalled by other members of the parrot family. Perhaps even more extraordinary is that some cockatoos have a very well developed sense of rhythm which enables them to dance to (but not imitate) the beat of music. Indeed, one such bird has become a You Tube phenomenon.

The enjoyment of music of Amazon parrots – or at least, certain types of music – is never in doubt as they chortle and display, flaring the tail and dilating the pupils in excitement. But then these species have very excitable personalities and loud and happy music stimulates them.

I remember the first and only live performance I saw of an Amazon singing in an operatic style. It was at San Diego Wild Animal Park, California, in the early 1980s. The parrot show presented by trainer Ray Berwick included an unforgettable act. Pancho was a Double Yellow-headed Amazon who, it was rumoured, had belonged to a soprano. His rendition, operatic style, of *I left my heart in San Francisco,* with the quavering cadences of an opera-trained voice, was so true and accurate as to be almost unbelievable.

I bought the single! Yes, Pancho had recorded this song with orchestration in 1981. Of course, these were the days of vinyl and this is the one record I regret losing over the years. On the flip side Pancho sang *When it's springtime in the Rockies* and *Bali Hi.*

Many people knew about Pancho and tried to teach their Amazons to copy him. You Tube is full of clips of quite pathetic attempts to imitate this incredible bird. I am sure that the fault lay with the teachers not the pupils!

However, much later, a second parrot diva hit the headlines in Colombia, South America. Called "Roberto Ruiz", it is claimed that this Yellow-crowned Amazon has become the first bird ever to record an entire song. (I cannot now recall whether Pancho also achieved this memory feat.) His rendition of the Mexican classic *La Cucaracha* (the cockroach) apparently made the charts. He recorded the song in a department store in Sucre, northern Colombia. It is claimed that after Roberto Ruiz made the recording, fans gathered outside the store to catch a glimpse of the maestro. That sounds like a public relations exercise to me!

17. PLAY

Corellas have a great sense of fun

It was sunset in Australia's Northern Territory. The orange sky gleamed through the woodlands as though they were on fire. My friends and I stopped by the highway to watch a flock of one hundred or more Little Corellas (Bare-eyed Cockatoos) feeding on the ground. The earth was sandy, with widely spaced gum trees and small patches of dark earth in which were growing clumps of grass. The cockies dug vigorously at the roots, probably pulling out little corms – perhaps of onion grass. They could not help interspersing play with feeding, one of a pair suddenly turning on its back and pulling its partner down with it. That could only be play for the sake of it, during the serious task of feeding before roosting! Cockatoos have a great sense of fun!

If you are ever fortunate to see a flock of Little Corellas, do not dismiss them as being too common to be worthy of observation! The chances are that they will make you laugh. In Perth, Western Australia, close to the Swan River, I watched a flock feeding in a large park. I was mesmerised! They were like a group of children, cavorting with their legs in the air, grabbing their companions' feet, swaying on grasses – and looking glad to be alive! At the slightest invitation to play, one would be on its back, rolling over and trying to pull another one down with it. The theory that play is just preparation for serious combat, holds no water here! Wild parrots can often be seen swinging from branches or, in the case of some Australia cockatoos, performing acrobatics from telephone wires.

The objects with which wild parrots can play are limited. These usually include siblings and other flock members, and items such as twigs. The most playful parrots include cockatoos, lorikeets and conures. Animal behaviour scientists state that only the more intelligent animals play and that these are the ones most capable of innovation. They say that play teaches empathy.

In 1994 I was travelling from Melbourne to Adelaide with friends when we heard a flock of Eastern Long-billed Corellas. They were playing and screaming on the roof of a large silo. We were compelled to investigate this joyful band of white clowns. Some of them were sliding down the steeply sloping roof – it was playtime before roosting! The sky was dull as they flew to nearby trees. Suddenly the sun came out and lit them with a golden glow, the sulphurous light of early evening.

Macaws are also playful birds. It is not unusual to see them hanging beneath branches, flapping their wings and screaming. In Bolivia Marc Boussekey, Jean Saint-Pie and Olivier Morvan observed a flock of fifteen Red-fronted Macaws perched in trees by a river, fluttering from perch to perch in animated play. Some were hanging upside down, wings wide open, on the thinnest of branches; others were bickering noisily. Such exuberance is surely just letting off steam – like children in a playground!

Duivenbode's Lory playing. They love to swing!

Play in aviaries

When my Duivenbode's Lories are swinging on their favourite playthings – a rope ball and a heart-shaped thick wire swing – I have to stop what I am doing to watch them. Sometimes they each hang on by one leg, simultaneously clutching the other's foot, or spinning round and round in a frenzy of excitement. The intensity of their play mesmerises me.

Or I see Green-naped Lorikeets *(Trichoglossus h.haematodus)* play-fighting on the ground, locked together, then tumbling over each other. One rolls on to its back, the other jumps on top, their beaks locked together for an instant. Then they tumble over, rolling, and one jumps up, whirring its wing to show the brilliant orange coverts. Then it is down again, beak and feet joined with those of its playmate. Finally after five minutes of rough and tumble, one falls on its back, feet in the air, as if to say: "I surrender!" Or I watch a little Iris Lorikeet *(Trichoglossus iris),* perched on a thin springy branch, then working the branch like a see-saw, so that it bounces up and down.

Swings are so simple to make and give so much pleasure to playful parrots. They do not need to incorporate a perch. A knotted rope that hangs from the roof, on which they can work up a speed, gives much enjoyment. However, any frayed ends in which they might catch their nails must be cut off.

Sadly, many parrots in aviaries have little opportunity to play. While some aviary birds do not use toys, partly because they have grown up without them, to others they are an immense source of enjoyment that keeps them entertained for hours. And that is what they need! (See **24. Environmental Enrichment**).

Highly intelligent parrots can play with humans in a different way – mind games, in fact! Large macaws are the masters of this! In the home they enjoy walking about and playing on the floor. They love to scare people by stalking them as though about to attack. For the most part they are bluffing – and enjoy seeing people turn tail and run! When sitting with someone they might suddenly let out a tremendous scream – just for effect. They know they can intimidate!

18. TERRITORIALITY

Parrots naturally defend the territory around their nests. Good sites are usually hard to find and must be protected from potential usurpers. Many parrots behave aggressively towards their own species only when the nest is threatened.

In Puerto Rico the Critically Endangered Puerto Rican Parrot is a prime example – nest sites are extremely scarce. In a typical territorial encounter, the owners of a nest will fly to the boundary of their territory, call excitedly and then engage in a battle with the intruding pair. "Generally the opponents grapple with each other, one on one, each bird lunging with its bill for its opponent's head or feet. Not infrequently two battling birds flutter to the ground locked in combat, giving high intensity vocalisations". Between bouts of fighting the pair members engage in intense mutual preening and occasionally bite chunks out of their perches (Snyder et al, 1987). You can imagine them attacking the branches in frustration at not being able to take a lump out of their adversaries!

So focussed are parrots when defending their nesting territory that they appear to be blind to danger and might even be preyed on by hawks – or their eggs or chicks predated in the nest which they have left unattended. On another Caribbean Island, that of St Vincent, the big handsome endemic *Amazona guildingii* is known to have been involved in fights in which two birds fell to the ground locked together – and were then caught by humans. Parrots are not unique in that the intensity of their battles endangers their lives. I have seen fighting Blackbirds locked together in my garden, oblivious of my close presence.

The aggressive behaviour of some parrots in the vicinity of their nests is the reason why comparatively few species in captivity can be bred on the colony system. Cockatiels, some Australian parakeets and Grey Parrots are among the few exceptions. Generally speaking, parrots that flock together are more tolerant and less territorial than those which occur only in pairs and family groups, such as the Hawk-headed Parrot.

Some parrot species are less territorial and nest in close proximity – even in the same tree. Examples in Australia include the *Polytelis* parakeets such as the Superb Parrot *(Polytelis swainsonii)*, called Barraband's Parakeet in Europe, and the closely related Princess Parrot (Princess of Wales Parakeet). Breeders enjoy these species for their gentle temperament. In a large aviary, more than one pair can be kept and bred without problems of aggression occurring – unlike other Australian parakeets. Regent Parrots (Rock Pebblers, *P. anthopeplus*) nest in loose colonies of up to 37 pairs, although most groups number only two to ten pairs. One study showed that colony sizes average 20 hectares (49 acres) but vary from less than one hectare up to 90 hectares (222 acres). Pairs normally nest in different trees, although two or more active nests have been found in one tree, which is unusual in the world of parrots.

In Western Australia a nest of the Red-capped Parrot (Pileated Parakeet) and the Port Lincoln

Parakeet *(Barnardius zonarius)* were located in the same tree, only 6m (20ft) apart. Only when both males visited the nest at the same time did the two males behave aggressively.

When nest sites are scarce, pairs that would not normally tolerate another pair breeding in close proximity waive the rule. In Costa Rica, many of the favoured nesting tree of the Great Green Macaw *(Ara ambiguus)*, the *almendro* (almond), have been selectively destroyed by man. It is the hard wood of choice to make floors, truck bodies and other items that tolerate hard ware. The result is that this magnificent macaw is on the Endangered list. Scientists studying this species were amazed to find, in 2011, three macaw nests in the same tree (see **29. Nest Sites**).

Cage territory

The owner of a Grey Parrot was tempted by a tame and adorable Black-headed Caique *(Pionites melanocephala)* in a pet shop. He was not prepared for a second bird. The shop owner generously offered to lend him a cage but this offer was unwisely declined. What did the purchaser do? He took the caique home and put it in his Grey's cage.

I found it shocking that a parrot owner could have so little understanding of a basic need of a parrot – its own space or territory. And shocking that he was unable to see the possible fatal consequences of his actions. Next day the caique was taken back to the shop. The Grey had attacked it and ripped out its tail feathers. The caique was lucky not to suffer serious injury.

A new bird must never be placed immediately with an intended companion but in an adjoining cage or aviary. In a new location it is at a psychological disadvantage, making it vulnerable to attack. After a few days of close observation the two birds can be placed together in an aviary that is neutral territory. If this is not possible, the original parrot should be introduced to the aviary of the new one (assuming that it is healthy), thus giving the new bird the psychological advantage. This is important because it will not be as confident as one long-established in the location.

In aviaries

The conscientious breeder will read all he or she can about the natural history of the species kept. This information can provide clues to breeding success. An example is Musschenbroek's Lorikeet from New Guinea. I had kept this species for some years before I realised that it was highly territorial. My pairs were within hearing but not in visual contact. Then I moved. I decided to place two pairs in a small unit of four aviaries, one pair on each side of a service area. This was a big mistake. Within a few weeks both females had denuded their breasts due to the stressful presence of another pair. One pair was moved but the breast feathers of both females never grew again.

Some keepers fail to understand how territorial most parrots are in the vicinity of their nest-box and when breeding. They might ignore the principal that you cannot safely introduce new birds into an aviary containing an established pair or group. They disregard the fact that certain species are dominant and aggressive and will inevitably attack more subordinate (usually smaller) ones. I am sorry to say that some zoos are the worst culprits in this respect.

To be set upon and killed by other parrots is a terrible death that zoos and owners must take every measure to avoid. A parrot's life can be lost in an instant because a keeper is careless. This can also happen in poorly maintained aviaries where a bird gnaws into the adjoining aviary and is at once killed by the occupants. It might also occur in a block of aviaries with inter-connecting doors. This type of construction is false economy! A row of aviaries should have a service passage from which each enclosure is entered by a door which does not connect to another.

19. ROOSTING

It was three-quarters of an hour before dusk on a very hot November evening. Biting insects were making themselves felt and the evocative laughing calls of Blue-fronted Amazons were filling the air. The parrots were coming in pairs, almost wing-tip to wing-tip, from every direction. They sailed down into their roost site, a group of large mango trees, calling excitedly and occasionally taking off again to circle around, shouting. Half an hour earlier a group of young people had been playing football on the grassy area in front of the trees.

I was in the Pantanal region of Brazil – the huge, seasonally flooded area as large as England. Staying at Caiman Lodge, I was bewitched by the Pantanal. Its open spaces allow one to view parrots in flight without the view being impeded by forest. Every day I was seeing eleven or twelve species of parrots, including the Amazons. Common bird! some of you might be saying. Not any more...

This Amazon's popularity, due to its ability to mimic and its beautiful plumage, means that removal of young from nests occurs on an alarming scale. This has been illegal in Brazil for decades but continues not only to supply the domestic demand but probably also helps (through illegal export) to fuel the trade in wild-caught Amazons in Argentina which is still legal.

The Amazons at their roost sounded so joyful; they radiated *joie de vivre.* They don't know it, of course, but they are the lucky ones. Here in the Caiman Ecological Refuge, all creatures are protected. No grasping hands come to snatch their chicks from the nest. The problem is worse than just taking away future breeding stock. Often the tree is cut down to access the chicks. Nest sites are at a premium.

I was there on two nights, to watch the parrots coming in and to hear the volume of their calls

These mango trees look very ordinary – a common sight in South America. But such large trees away from forested areas are invaluable as roost sites – here for the Blue-fronted Amazons at the Caiman refuge.

building to a crescendo. I was happy to see them enjoying life so much, knowing they were safe forever and that their breeding successes would help to preserve this once common species.

One morning I saw distant pairs leave their roosts at first light, hearing their chuckling calls as they flew overhead. Some pairs perched in nearby trees and preened each other before they took off for their feeding grounds. This roost is the centre of their world. More needs to be done to protect such sites.

Various sites

The most common roosting sites for parrots in general are, of course, tree branches but cavities in trees, cliffs and rocks, termitaria and the interior of epiphytic plants are also used. In Africa, certain lovebirds roost in weavers' nests. Some parrots prefer to roost in their nest site, thus preventing other birds from taking it over, but others leave the area when the young have fledged and might be hundreds of miles from their nest. The downside of roosting inside cavities is that if a nocturnal predator enters, there is no escape unless the site has two exits. This is unlikely, except in the case of lovebirds that nest in mopane trees that have multiple cavities.

In Guyana, the Endangered Sun Conure *(Aratinga solstitialis)* has different roosting habits that almost led to it being trapped to extinction. In 2009 Toa Kyle arrived in the village of Karasabai, on behalf of the World Parrot Trust, to study one of the last known populations. In the 1980s local people caught large numbers and sold them to dealers from Georgetown. Then a dealer pulled a gun on the villagers to avoid paying so they immediately made the decision to stop trapping. The gun-happy dealer inadvertently saved this population.

Toa Kyle discovered that Sun Conures roost in tree cavities throughout the year. He found three roosting sites, two of which were in conspicuous, open locations. By placing nets over the entrances trappers could capture an entire flock and wipe out the local population. In less than thirty years this beautiful parakeet was almost extinct (Kyle, 2009).

Most parrots sleep with the head turned over the shoulder, with the face and beak pushed into the feathers of the back. Abyssinian (Black-winged) and Red-faced Lovebirds *(Agapornis taranta* and *A. pullarius*) and Hanging Parrots *(Loriculus)* often sleep upside down, well disguised among leaves. If disturbed by a predator they simply drop down, open their wings and, hopefully, are gone before it strikes.

Communal roosting

Many parrots roost communally. One reason why communal roost sites occur is because there is safety in numbers. A nocturnal predator would have more chance of being detected and the risk to each individual parrot would be lower. Such gatherings are probably also a matter of convenience as in some areas there are insufficient sites for large numbers of parrots to roost in pairs in different places. Roosts could also be the equivalent of human clubs, where birds can meet a partner!

For most species there would not be enough cavities in which to roost and even if there were, some parrots prefer not to use their nest site for roosting. In the Serra Geral escarpment of Rio Grande do Sul state in Brazil, a pair of Maroon-bellied Conures *(Pyrrhura frontalis)* roosted in a primary branch of a *Trichilia claussenii* tree. This cavity was later taken over by a pair of Purple-bellied Parrots *(Triclaria malachitacea).* When their nesting attempt failed the cavity was used for roosting on alternate days (they apparently took it in turns!) by conures (perhaps the same two) and a pair of Short-tailed Antthrushes. Purple-bellied Parrots do not roost in cavities. These habits persist in captivity. *Triclaria* never roost in nest-boxes; *Pyrrhuras* always do.

South and Central America

Among the few species still numerous enough to sleep in congregations of hundreds, or even thousands, are the Budgerigar and the Orange-winged Amazon. Not far from Guyana's capital, Georgetown, there is such an Amazon roost, in large clumps of bamboo. Another notable site is that in Guajara Bay, south of Belem, in Brazil, on the 7.4 hectare (18 acre) Ilha dos Papagaios (Parrot Island). Densely forested, a twice weekly count was made there by three teams of researchers, who were careful not to duplicate parrots counted by another team.

Occupancy in 2005 varied according to the month, with June averaging 7,084, and July 8,539. The highest count was 9,603 on July 2 when the young were independent from their parents. The lowest counts were between October (1,545) and December (1,364). Pairs began to leave between August and October and returned with young in January. The island is an important refuge for the parrots in an area where poaching is otherwise intensive (De Moura, Vielliard and Da Silva, 2010).

For me, watching a congregation of parrots leave a roost in an area that is not densely vegetated (thus giving a clear view) is one of nature's most thrilling spectacles. Sometimes it is hard work to achieve it! In 2005 I was fortunate to spend a few days in the Colombian Andes with the young and dedicated ornithologists of ProAves (the leading Colombian bird conservation organisation). The focus of our attention was the charismatic and so-nearly-lost-forever Endangered Yellow-eared Parrot *(Ognorhynchus icterotis).* One of the few known roosting areas is above Jardin, a delightful little town in the Western Cordillera in the department of Antioquia.

After driving for an hour, ever higher in the Andes, we climbed for two and a half hours on steep, rocky paths, scrambled across dried mud marked with deep rivets and made our way through swamp and across streams. When we reached the top, as the sun was going down, we found a denuded landscape. The Critically Endangered Quindio wax palms *(Ceroxylum quindiuense),* on which the parrots rely for food and nest and roosting sites, had nearly all gone. The pastures were so poor the landowners cut trees in order to survive.

Yellow-eared Parrots became extinct in Ecuador during the last years of the 20th century, and were believed to have suffered the same fate in Colombia. Then one wonderful day in April 1999 a flock of sixty-one birds was discovered. Now here I was, watching a pair of Yellow-eared Parrots fly into the top of a dead wax palm. They were the lucky ones who had claimed their own hole. Thirty-three more parrots arrived, little more than silhouettes in the fading light. They headed for two palms located about 15m (50ft) apart. Here they roosted in the fronds.

Not far away seventy or so roosted in the second palm. I felt deeply privileged to be seeing them. The parrot had hung on to survival in a remote area where local people had no idea they were the last of their kind on the planet. All that soon changed! A robust conservation programme turned around the fortunes of the Yellow-eared Parrots so rapidly that today there are more than 1,400. In 2010 the species was downgraded from Critically Endangered to Endangered – an almost unheard of improvement in the fortunes of a parrot species.

But on that day in February 2005 fifty Yellow-eared Parrots spent the night in a palm tree just below the *finca* where I slept on the floor. It was the darkest night I have ever known, without electricity, but below a sky which was a solid, brilliant mass of twinkling stars.

Next morning we started the difficult downward climb before first light. We had to reach the roost trees before the parrots departed! Already a few were moving about in the fronds. They started to chatter and the synchronisation of their voices indicated that they were getting ready. Then, with a whoosh of wings, the entire flock was gone. Gone and out of sight in ten seconds!

In north-western Costa Rica, in the Guanacaste Conservation Area, scientists from Cornell Laboratory of Ornithology studied Orange-chinned Parakeets *(Brotogeris jugularis)*, Orange-fronted Conures *(Aratinga canicularis)*, White-fronted Amazons *(Amazona albifrons)* and Yellow-naped Amazons. All four species used communal roosts. The only one with a fixed site was the Yellow-naped Amazon – a place that had been used for more than 30 years.

The White-fronted Amazons and the Orange-chinned Parakeets would use a roost for three to six weeks and then move on, usually many kilometres distant. The Orange-fronted Conures changed their roosting site every night, possibly because they are a favourite prey of a large carnivorous bat, *Vampyrum spectrum.* The three species that changed their nocturnal locations spent the last two hours before roosting at "staging" sites, vocally advertising the proposed place to recruit passing birds. The researchers found that: "Often several nearby but competing sites are advertised simultaneously until enough defectors from one site cause all the others to give up and join the main mass" (Bradbury and Balsby, 2006).

The staging often occurred in trees near to, but never in, the final sleeping location. At the right time, the parrots would move into the denser foliage of the chosen site. However, any of the three species might suddenly explode out of the tree, wheel around the sky for several minutes, then begin staging and roosting all over again.

In Argentina, at the famous El Cóndor breeding site of Patagonian Conures *(Cyanoliseus patagonus),* pairs roost in cliffs facing the Atlantic ocean. In the cliff-top habitat where trees are scarce or stunted, these handsome parrots perch on telephone lines, one hundred or more together, sometimes dropping down to the scrubby ground to walk about in search of food. In the evening the non-breeding birds gather in their hundreds against a backdrop of the oldest lighthouse in Patagonia and the evening sun turning the sky orange.

Then they fly to roost to the little town where telephone lines are favoured sites. One roosting area was near the house in which I stayed for a few memorable days. I could hear their strident calls throughout the night! What a pleasure that was!

Above, Patagonian Conures gathering to roost on telephone lines. Below, also known as Burrowing Parrots.

Pacific Region

In Australia, Eastern Rosellas roost in groups of twelve to twenty-five birds, in eucalyptus trees and saplings and in *melaleuca* thickets. They land in the smaller branches and climb into the outer twigs where they hide in the dense mass of leaves. Their preference for certain trees is very strong. If they are disturbed, they fly out and quickly return.

In Indonesia, the gorgeous, once common, Red and Blue Lory *(Eos histrio)* is classified as Endangered with a total population of fewer than four thousand individuals. Trapping and habitat loss have drastically reduced its numbers. Found only in the small islands of the Sangihe-Talaud group, this includes Karakelang, the largest of the Talaud Islands in Indonesia.

In 2013 Mehd Halaouate was taken to a roosting tree by Michael Wangko from Kompak. This is a local conservation group collaborating with the World Parrot Trust to try to end the illegal capture and smuggling of the lory. The fig tree *(Ficus variegata),* believed to be the most important roosting tree on the island, was about 45m (147ft) high with a crown 7m (23ft) wide. The trunk was smooth and without branches until near the top.

The lories started to arrive at 5.35pm but did not land on the roost tree until 5.50. By 6.10pm their chatter was deafening. Some did not stay and left in total darkness! "We could hear their wings flapping, and could see dark silhouettes flying away thanks to the light of a full moon" (Halaouate, 2014).

The chatter of the lories continued until 8pm. At 4.15am they were awake and starting to converse. At 4.28 a few were flying between the branches and at 5.34 the first birds started to leave the tree, most in pairs but some in small groups. By 6.05am all– an estimated five hundred plus – had departed. How vulnerable a species is to trapping when a large part of the population uses one roost! And how important large trees are to their survival.

In cage or aviary

Caged birds have the choice of roosting on a perch or clinging to the wire. They might choose whichever point is highest. If the cage is almost bare of "furniture" they might roost next to the only toy – a Budgerigar right under its bell. Presumably closeness to such items give some sense of security, indicating that a totally bare cage is not desirable. Some owners provide small cloth tents made specially for parrots. However, if the fabric is not suitable there is a risk that a bird could catch its nail in it, possibly with disastrous results. This has happened. The interior of such an item should be inspected before purchase.

In the wild some parrots roost in their nest site throughout the year. These species, including lories and small neotropical parrots (caiques and Hawk-headed Parrots) will always roost in their nest-box if this is permanently in position. They thus have a cosy and safe roosting site – but the downside is that they may nest at inconvenient times of the year.

Some parrots like to rest on a flat surface – just as we do. The alternative to providing a nest-box is a shelf with two nails in the back that hook onto the side of the cage. This simple piece of furniture is especially appreciated by older birds that might find it difficult to grip, due to arthritis.

Parrots in outdoor aviaries need extra care. Unfortunately, many die because they are not shut into secure quarters at night and are vulnerable to nocturnal predators, such as owls, rats and stoats. In cold climates, Ringnecks and other *Psittacula* species are extremely susceptible to frost bite, resulting in the painful loss of toes and even feet. If the condition is so serious that the blood supply to the toes or feet ceases, those parts will become black, shrivel up and fall off. The parrot will be a cripple for the rest of its life.

I feel it is the responsibility of every carer of aviary birds to ensure their nocturnal safety. Shutting them inside at night is probably the most important factor in preventing premature deaths. I despair when I hear of birds being killed in their nest-boxes by rats and stoats in outside flights. An aviary that is not vermin-proof is not fit for purpose. (I am not including mice, which can be extremely difficult to deter due to the small size of young ones.)

Owners of companion birds should be aware that caiques and conures love to roost in cavities and might try to enter small spaces between furniture or at the back of fires, for example. Any such holes should be blocked up before they bring their bird home.

If you are in the south of England and you want to observe a parakeet roost site, go to Wormwood Scrubs park in Hammersmith, West London. More than one thousand naturalised Indian Ringneck Parakeets roost in the trees there. They gather in pre-roost sites then, as the numbers increase, they fly to the main roost. The same thing happens at Mitcham Common, not far from Croydon. The whole area is alive with them and other roosts occur in Ewell, Epsom and Reigate.

20. LESSER-KNOWN SENSES

We already know **(5. The Eyes have it!)** that parrots have excellent vision. Their sense of taste is well developed in comparison with many other groups of birds. Their sense of touch is extremely good: tongue, feet and beak are frequently used. Their hearing is excellent. I suspect it is better than ours. When you are very close to some species, you can hear that they make low-pitched sounds that are hardly audible to the human ear. But we do not know a great deal about their other senses.

It has long been known that animals, including birds, can sense imminent tectonic activity. Dogs will start to howl and wild birds behave abnormally, ceasing to sing, for example. Indeed, in 2014 scientists were planning to fit transmitters to birds in disaster-prone areas because their behaviour can give a warning of impending catastrophes that precede by some hours any scientific equipment used for this purpose.

Parrots are aware of imminent earthquakes. The owner of a Timneh Grey Parrot called Jing showed that she could predict earthquakes after the disastrous event that hit Los Angles in 1994. Jane Hallander and Jing lived 720km (450 miles) from the epicentre of this quake, yet eight hours before it happened Jing was staring at the ground, as if in a trance, or hanging upside down from the perch. That night she did not want to enter her cage. Next morning she hurried out of the cage and resumed the staring behaviour for the next few hours. Her owner could subsequently predict the magnitude of the coming earthquake by the intensity of Jing's behaviour.

Birds are more susceptible to subtle environmental changes than are humans. Animals are more sensitive to the ultrasonic waves that precede an earthquake or they may sense the chemical or electrical changes in the field of the earth which could be indicators of earthquakes. Japanese physicist Professor Motji Ikeya believes that wild birds are sensitive to electro-magnetic pulses along an earthquake fault line and sense them when tiny cracks appear in pressurised rock, creating electric currents.

Jane Hallander's Timneh is unlikely to be unusual – Jane herself was unusual in reporting what she saw, as was an owner of a pair of Grey Parrots. She was awoken at 3.59am by the birds crashing against the side of their cage. An earthquake hit at 4.04am.

Parrots (and other birds) are like barometers: much more sensitive to changes in atmospheric pressure than are humans. They know before us that a storm is coming. The seven Greys of my friend, who spend the daytime in an outdoor aviary, are keen to be brought inside two or three hours before a summer storm hits.

Sense of smell

Various experiments with birds have shown that generally the sense of smell is not well developed in birds, compared with that of mammals.

Exceptions are a few species with highly specialised dietary requirements, such as Kiwis and Albatrosses. It is known that in parrots the olfactory lobes (the part of the brain that deals with smell) are small. This is not surprising as they do not locate their food by smell.

T.J.Roper of the University of Sussex's School of Biological Sciences carried out possibly the first behavioural investigation of olfactory capability in any parrot species. His subjects were two hand-reared Yellow-backed Lories *(Lorius garrulus flavopalliatus).* He chose this species because it feeds on nectar, flowers and fruit and might use the sense of smell for foraging. The lories had to do this to distinguish a container of water from one of nectar. In each of the experiments tubes of cotton wool soaked with one of three essential oils (palmarosa, geranium or patchouli) were taped to the container of nectar. The study showed that the lories appeared to discriminate a dispenser containing artificial nectar from one containing water, using their sense of smell (Roper, T.J., 2003).

In New Zealand researchers worked with Keas and Kakas *(Nestor meridionalis)* to assess whether they displayed varying behavioural responses to different types and concentrations of scent. Video monitoring was used to measure parrot visits to scent stations compared with controls and to assess any tendency to explore novel odours. Although sample sizes were small and individual responses varied, both species showed an ability to distinguish between scents and controls, and to detect novel scents (Gsell *et al,* 2012).

What significance does this have for the parrot keeper? Not a lot – except the knowledge that different odours are detected. A friend told me that her macaw knew when she was cooking bacon – which he relished – because he would call out for some. Perhaps he could smell the bacon cooking – but the possibility exists that the clatter of pans or some other auditory cue alerted him.

Feathers of the Black-winged Lory have a strong odour.

On the subject of odours, many parrots have a faint but pleasant smell – not connected with the conditions (clean or dirty) in which they are kept, but peculiar to their species. What interests me is why one particular species of lory, the Black-winged *(Eos cyanogenia)* has such a strong smell, which is most apparent when in breeding condition. I would like to know how and why this odour is produced. The closely related Red Lory *(Eos bornea),* for example, has only a faint odour.

Thought transference

If you have a close relationship with a parrot is thought transference a possibility? In 1888 the first volume of W.T.Greene's classic work *Parrots in Captivity* was published in London. He quoted a correspondent who wrote to him about a female Alexandrine Parakeet *(Psittacula eupatria).* "I believe that her fondness for, and her sympathetic attachment to me, was something more than mere instinct for if I think strangely of her at any time, even in the middle of the night, she is sure to answer me with her own little note, her eyes remaining shut, and her head tucked in her shoulder, as though she were fast asleep."

It seems unlikely that the correspondent invented this for 120 years ago the subject of thought transference between human and animal would hardly have been considered.

Now let me mention another classic – one of my favourite books. *Birds as Individuals* was published by Len Howard (real name Olive Howard) in 1953. The author had a unique relationship with wild birds, especially Great Tits. She wrote about a female who seemed to be able to communicate with her mate – or her hearing was extraordinarily sensitive. Sometimes when she was inside Len Howard's cottage she would suddenly stand rigid, with a tense expression, dropping the food she was eating tucked between her toes. Then she would hurriedly fly out of the window and across the meadows to the copse. Sometimes Len Howard followed her to find that she had flown to her mate. 'The sudden tension of her body and her facial expression, as if vision was centred inwards, her eyes not focusing upon what was in front of them, suggested she was responding to some form of communication from her mate.'

In **Intelligence (14)** I suggested that the degree of intelligence shown by a tiny bird like a Great Tit must surely be exceeded in a parrot. Likewise, the experience related by Len Howard regarding possible thought transference between Great Tits surely means that this ability is also present in parrots and possibly to a greater degree.

Unfortunately, the New York owner of a Grey Parrot called N'kisi made claims about his telepathic abilities that made thinking people sceptical or downright sarcastic. One comment was that the title of the paper "Testing a Language-Using Parrot for Telepathy" would send most journal editors to their grave, killed by laughter! I believe that telepathic communication can exist between two birds or a bird and a human companion, but the experiments in which N'kisi participated were not convincing in the way they were set up or in the interpretation of the outcomes.

PART IV.
WHAT PARROTS NEED

Red-collared Lorikeets gather to drink at a dripping tap.
From a photograph by the author.

21. THE BASICS: CLEAN WATER, AIR AND FOOD AND SUNSHINE

The needs of parrots are the same as those of other birds: food, water and shelter – and more so than most species, the constant company of their own kind.

Clean water

Most parrots drink water by scooping it up with the tongue, and pressing the tongue against the palate to force the water back. The daily intake needed has been calculated at 2.4% of the body weight – but this varies according to species, temperature, habitat and food type.

Most parrots need to drink at least once a day. This is why there are no parrots in the deserts of Western Australia. With the exception of Antarctica, Australia is the driest continent, with an average rainfall of below 600mm per annum.

The Galah is a cockatoo found in arid regions. Experiments on the temperature regulation and water economy of the Galah were carried out by scientists with captive birds (average body weight 270g), with a few measurements from wild birds. It indicated that body temperature varied from 38.2°C (at night) to 41°C (or just above) during the day. Under moderate temperate and humidity conditions, they need to drink 7.3g of water every 24 hours, at the same time taking about 9g of food (which consisted almost entirely of sunflower seed – highly inappropriate for this species). Without water they lost 2.2% of body weight daily. After ten days they had reached the tolerance of water deprivation and sat quietly with eyes closed, and were given water *ad lib.* (Dawson and Fisher, 1982). (I wonder what the researchers would have looked like after being deprived of fluids for ten days!)

Some wild parrots descend to the ground to drink, especially in arid areas where water is hard to find. They are then vulnerable to predators so they prefer to go down in a flock, as this reduces the individual's chance of predation.

In Australia surface water is limited and throughout arid and semi-arid areas many birds drink from cattle troughs. In 1988, a friend thoughtfully set up a photo opportunity in the outback of New South Wales. He placed a hide in a windmill overlooking two water tanks. The hide was about 2.4m (8ft) above the ground and perfectly situated. Sunset would be at 8pm and we went inside at 6.30pm and settled down to what I imagined would be a long wait. However, very soon a pair of Galahs flew in, followed within minutes by a pair of magnificent pink and white Major Mitchell's Cockatoos. By 8pm we had seen twenty-four of each species drinking at the tank. What interested me was that the cockatoos took the fresh water as it spilled out from a pipe before it entered the tank. In contrast kangaroos, crows and pigeons were drinking at a sheep trough. It emphasised that parrots chose clean water when it is available.

Little Corellas seeking water in Perth

A decade later, in northern Australia, Red-collared Lorikeets lit up the trees wherever I went. Their rich colours glowed in the intense sunlight against the backdrop of a cloudless cerulean sky. With the blue head, orange breast and beak, and green wings, no other birds in the region could compete with this array of colour. In flight, the orange and yellow feathers of the under wing coverts were dazzling.

One particular observation lives in my memory. A group of lorikeets descended from the large tree in which they were noisily perched to drink from a dripping tap (see illustration on page 103). Two or three perched on it while others waited their turn on the ground below. Lorikeets are, of course, nectar feeders, but they still need the plainest liquid of all.

Some parrots do not use traditional water sources if a convenient alternative is nearer. In South Australia Barnard's Parakeets, also known as Mallee Ringneck Parrots, were seen following a water truck that was spraying water on a road prior to resurfacing. Where small pools of water settled, the parakeets flew down and took a drink.

It is interesting that I never see my own lories and lorikeets drinking water. They are enthusiastic bathers and will, on occasions, bathe in water that is not fresh, but more often they avoid it and wait until I have changed the water – a daily task.

I once came back from a few days away to find my birds' drinking water containers in the inside accommodation filled nearly to the brim. It is not possible to put full stainless steel containers inside the ring that holds them without spillage. It seemed that the temporary carer had refilled the containers by pouring fresh water on top of the old. Here was someone who did not understand that clean water containers are equally as important as clean water. Or perhaps the reason was laziness! Whatever, my advice is to use stainless steel because it is easily cleaned and cannot crack and hold bacteria.

It is worth noting that not all parrots drink water throughout the year. For example, in the Pantanal, the Hyacinthine Macaws drink the liquid in unripe palm fruits, by holding the fruit in the beak and tipping the head back. I have seen captive Scarlet Macaws take a slice of orange in the beak, tip the head back and drink the juice by crushing the fruit.

Some parrots obtain moisture from plants. In Australia, the Critically Endangered Orange-bellied Parrots *(Neophema chrysogaster)* obtain water from the succulent leaves of salt marsh plants. In South Africa, the Endangered Cape Parrots *(Poicephalus robustus)* also take their moisture from plants. They have been seen drinking dew drops from Spanish moss hanging from emergent branches in thick mist. In the neotropics, many parrots that live in the canopy drink the water that collects in bromeliads.

Clean air

There are city dwelling parrots throughout the tropics and in more temperate climates, such as the renowned naturalised populations of Amazons and conures in Los Angeles and Ringneck Parakeets in a number of European cities, including the UK. If and how they respond to air pollution is not known. However, they do have a choice regarding where they live.

This is not so regarding our parrots. Unfortunately, some of them dwell in rooms or houses where the air is badly polluted with agents detrimental to parrots, such as air fresheners and cleaning products that give off harmful fumes. (So do most paints, so eco-paints should be used, if possible.) Unlike humans, parrots do not have expanding lungs. They have air sacs that can take in air but cannot filter out poisons. Their respiratory system is faster than ours and they need to take in much more oxygen than we do. Note that candles can use up the oxygen in the air, killing birds in the process.

Because their air sacs cannot filter out poisons they are extremely susceptible to airborne pollutants. One of the worst is cigarette smoke, a dangerous cocktail of about 4,000 chemicals, according to Cancer Research. These include more than 70 cancer-causing chemicals such as tar, arsenic, benzene (an industrial solvent), formaldehyde and acrolein which was formerly used as a chemical weapon. Nicotine on hands and fingers is also harmful. So – if you are a smoker and you have just read this, do you care about your parrot? If not, carry on smoking and find your parrot a better home.

Smokers are usually reluctant to accept that this habit is harmful. Two Black-headed Caiques belonging to one woman died at a young age – not kept simultaneously – from a "heart attack". That was her diagnosis. Then she acquired a macaw, which also died after several years. When she acquired a young Blue and Yellow Macaw I felt obliged to reiterate my warning about smoking in the home. Her response was that she knew of a parrot that had lived in a pub for years.

Well – perhaps it was the same one that veterinarian Alan Jones wrote about in his book *Keeping Parrots – Understanding their Care and Breeding.* He related the story of a Double Yellow-headed Amazon, a wonderful character called Popeye with an extensive vocabulary of words and sounds. After five years with his caring lady owner, one day he fell off the perch and died. Alan Jones carried out the post-mortem examination.

He wrote: "All his air-sac membranes (which should be thin and transparent) were thickened and cloudy, and dotted with black spots of soot. The lungs were congested, and also filled with black spots. This is known as anthracosis, and is the result of accumulation of hydrocarbon particles from cigarette smoke in the respiratory system. In addition, the major vessels leading from the heart were yellow and thickened with fatty deposits, known as atherosclerosis. This obviously had the effect of reducing the diameter and elasticity of these arteries, thus increasing the load on the heart.

"This pathology is the direct result of the inhalation of tobacco smoke, and Popeye ultimately died because of the long-term damage sustained in this way."

Popeye's family were devastated. They were all non-smokers Researching his past revealed that Popeye had spent thirty years in a pub, in the public bar. By the time the family received him, the damage was already done, with fatal consequences.

Carbon monoxide poisoning

The deaths of 730 wild-caught Grey Parrots one Christmas Eve, on a flight from Johannesburg to Durban, made headlines all over South Africa and sent shock waves throughout the bird-keeping world. When the crates were unloaded, 730 were dead and ten died later. So the entire consignment perished.

The parrots were part of an order of 1,650 adult Grey Parrots (the rest were still in quarantine) which were caught in the Democratic Republic of Congo to be sold to South African dealers and breeders. They were imported with valid CITES permits. This shipment was not illegal. I will not delve into the ethics of the trade in wild-caught parrots here – except to say that it is abhorrent (see **38. The Wrongs heaped on Parrots**).

This terrible story raises a lot of issues, one of which could be relevant to everyone who keeps a parrot. The unfortunate Greys were believed to have died from carbon monoxide poisoning. A small dog on the same flight survived – but then, as already mentioned, birds are so much more sensitive to gases and fumes than mammals, because their respiratory system is quite different.

Carbon monoxide is an odourless, tasteless and colourless gas which has been called the silent killer. It kills people as well as birds. A beloved Grey Parrot died as the result of inhaling fumes from a gas fire. The owner escaped with his life. Symptoms in humans are dizziness, nausea and fatigue and it seems likely that birds suffer the same – but they react more quickly than humans.

I recall a distressing story from some years back regarding the deaths of two rare Amazon parrots that were transported in the boot of a car. They were dead on arrival. I thought most bird keepers knew that birds in transit should never, ever, be placed in the boot where they are at high risk from carbon monoxide poisoning. But I was wrong. I repeat the warning here. Birds should never travel inside the boot of a car.

It is not always realised that carbon monoxide poisoning does not result only from badly installed appliances that burn natural gas, but also from kerosene and other fuels and from generators that might be used during a power cut.

Everyone who uses these fuels or who keeps a parrot in the house, is advised to buy a carbon monoxide detector. It is not expensive. There are three types: those that plug into electrical outlets (not recommended where young children are present), battery operated detectors and those that are wired into the electrical system. The latter require installation by an electrician. Inexpensive detectors can be obtained from do-it-yourself and hardware stores (look in the smoke alarm section). They produce an audible alarm.

The detector should be not be placed close to an outside door and at least one metre away from appliances such as boilers, heaters and cookers. Because carbon monoxide (CO) gas rises, detectors are best placed at about head height – but not on the ceiling.

Carbon monoxide is especially dangerous to birds in poorly ventilated, heated areas. When CO binds to haemoglobin in the blood, it results in hypoxaemia, that is, inadequate oxygen in the blood. Death is swift unless the bird can very quickly be moved into fresh air. According to *Avian Medicine, Principles and Application,* "emergency care should include the administration of 90 to 95% oxygen in a cool, dark, stress-free environment." However, death can happen so quickly that there is seldom time to act.

Sunshine

Most parrots originate from the tropics and sub-tropics so there is a belief that these birds love to sit in the sun. Not true! Many parrots are canopy-dwellers and spend much time amid dense vegetation, where their bright colours are not obvious. There are exceptions, but few parrots indulge in sun-bathing. The most notable exceptions must be the Vasa Parrots *(Coracopsis).*

They sun-bathe like pigeons, even lying on their side on the ground, assuming exaggerated postures, usually with one wing outstretched and tail fanned to one side. To the uninitiated this looks like a dying bird! Zoos need to put a warning notice on the aviary to qualm the fears of concerned visitors!

In our northern climes, aviary parrots avoid hot sun but enjoy early morning and evening sunshine which often stimulates them to bathe, as does sunshine streaming through a window. However, the beneficial ultra-violet wavelengths are filtered out through glass. Research on humans in the UK suggests that exposure to UV light (sunshine or tanning lamps) lowers blood pressure, thus reducing the risk of cardiovascular disease (heart attacks and strokes). It seems likely the same applies to parrots. Many companion birds seldom enjoy exposure to ultra-violet wavelengths – and many die from cardiovascular disease.

Companion parrots that are kept permanently indoors will obtain great benefit from short periods outdoors, weather permitting. The best way to achieve this is to take a parrot outside in its cage, or in a travel cage that allows sunshine to penetrate. The cage door should be padlocked! It should never be left unattended because of the risk of harm from cats and hawks. If a parrot is unfamiliar with the outdoors it will probably seem nervous at first, phased by the great expanse of sky, and will turn its head to one side and look upwards. It will need reassurance!

Never forget that to leave a parrot in hot sun is cruel. It must have access to shade. The same is true in aviaries: a section of the roof must be covered. Parrots kept indoors should never be placed in a conservatory unless it has blinds on the roof and sides. I know of a Sulphur-crested Cockatoo that died due to the owner's neglect in leaving it in a conservatory on a very hot day. This is the equivalent of leaving a dog in a sweltering car with no windows open.

Greater Vasa Parrots (Coracopsis vasa) sunbathing.

It should be noted that Vitamin D deficiencies are common in captive birds kept indoors in a UV deficient environment with insufficient dietary vitamin D. Psittacine birds are commonly fed a seed-based diet; seed is deficient in this vitamin. Those kept indoors receive inadequate UV for cutaneous synthesis. It is well established that Grey Parrots are particularly susceptible to hypocalcaemia. It is suggested by Michael Stanford, a veterinarian from Cheshire, that seed-based diets, which are deficient in vitamin D, are a contributory factor to this condition – a subject that he studied extensively.

Gnawing wood

An urgent need of parrots which is sometimes overlooked is **The Need to Gnaw.** Please refer to this section **(22)**.

22. THE NEED TO GNAW

Parrots might be called the rodents of the bird world in that they have a great need to gnaw. In the wild this is fulfilled by biting at the bark on tree trunks and branches, visiting clay licks to remove chunks of mineral-laden earth, biting off the hard outer shell of fruits and nuts and gnawing at the wood around the nest entrance. In a few cases, they gnaw at the sandstone cliffs in which they nest.

In captive birds, the motivation to gnaw at wood or destroy cardboard boxes, for example, is very strong. Most parrots that are not afraid of new objects introduced into the cage or aviary will almost immediately start to bite at anything gnawable! There is no hesitation. Gnawing is a need that is as strong in a parrot as it is in a rodent! Pity the parrot that sits in its cage with nothing on which to exercise its beak! The cage is metal, the food pots are steel or plastic. How frustrating that it cannot use its beak! The next step is over-preening its feathers, leading to feather destructive behaviour.

For birds in large cages and aviaries, fresh-cut branches with leaves should be offered regularly – at least once a week. They make an ideal starting point for some creative foraging ideas on the part of the owner. But even on their own, what can be more natural for tree-dwelling birds?

One of my best memories from the time I was Curator of Birds at Loro Parque was of the days when all the parrots were given fresh branches of pine or *Casuarina equisetifolia,* which arrived by the lorry load. The parrots were so excited when they saw the branches coming! These were soon reduced to a pile of wood chippings. Such branches are especially important for fig parrots and caiques who tend to suffer from overgrown beaks.

Usually the first action of any parrot is to remove the bark. This is instinctive and helps to fulfil that great need. It is an important part of a parrot's life that should never be overlooked. Yet often it is!

In Europe the best trees to use are apple or the wood of other fruit trees, or hazel and willow. Willow is acceptable for enrichment but unless branches are very large it is not really hard enough to keep a parrot's beak worn down. In Australia and other warm climates eucalyptus and Casuarina are greatly enjoyed for enrichment, perching and gnawing. Galahs will use eucalyptus leaves to take into their nests.

It is a lot of work to keep aviaries and cages supplied with suitable branches. Most parrots would be nibbling and gnawing nearly all day long if provided with enough fresh-cut wood or berry-laden branches such as hawthorn. Finding a supply can be a major problem for some urban dwellers. It is a good idea to seek advice from local parrot keepers who might be able to recommend suitable trees.

Branches have another beneficial aspect. Tree bark appears to contain elements that are important or nutritious. It is readily eaten or destroyed by nearly all parrots who tear at the bark as soon as the branches are offered. In the wild Severe Macaws, for example, have been seen to eat the bark of the cotton tree *(Ceiba pentrandra).*

On Vancouver Island in Canada Wendy Huntbatch founded and runs the World Parrot Refuge. Volunteers spend countless hours making gnawable toys for 700 plus relinquished parrots and cockatoos – but for Christmas 2013 she had a brilliant idea. (She is full of them!) She e-mailed me:

"We have just had a FUN raiser for the parrots. Usually people take their used Christmas trees to a recycling unit or wood chipper. I thought the parrots would love them so I contacted the local papers. They published my request and we have received hundreds of pine trees. Right now the Refuge is looking like a forest! One of the local car dealers offered their services to pick up and deliver for folks who don't have suitable transport. Incidentally, these trees are not treated with a fire retardant. There are many tree farms around here and buyers just go and cut their own."

So to the people who have asked me if Christmas trees can be used for this purpose, I would say,

Gnawing fresh-cut wood is vitally important for all parrots, including Meyer's **(Poicephalus meyeri).**

check with the supplier to find out if they have been treated with any chemicals.

The destructive urges of two escaped Amazon parrots made headlines around the Yorkshire area of Cottingham, in the UK. They had lived through several bitterly cold winters at their headquarters, the church of St Mary the Virgin. In 2008 they were the prime suspects in the disappearance of gold leaf from the clock face of the church. It must have been irresistible to their busy beaks!

23. THE ELEMENTS

Rain and humidity

The exuberant joy with which wild parrots bathe in the rain is always a delight to observe. In a leafless tree they will spread their wings and ruffle their feathers, perhaps even hanging upside down from the branch. In so doing they capture moisture on the underside of the wings and body as well as being drenched on the upperparts. In leafy trees parrots have even more fun and flutter through the wet foliage in a frenzied manner. Wet plumage results in a prolonged preening session which helps to maintain the feathers in the condition needed for flight.

Rain is also a signal to breed for species in arid environments, for it will be followed by green leaves and shoots and an increase in the variety and quantity of food available. In parts of Australia, parrots might not breed at all in a year in which there is little rain.

Some captive parrots, such as white cockatoos *(Cacatua species),* can be stimulated to breed with the use of a sprinkler set on the roof of the aviary. After a prolonged dry spell, turning this on for a couple of times daily acts like an aphrodisiac! Providing good bathing facilities is very important. Water containers might not be large enough to bathe in or they are wrongly situated.

Drinking water is often provided in one of a set of swing feeders which means that if birds bathe in them, the water will be dispersed into the food dishes. Baths need to be large – at least as long as the body length of the parrot – and they should be situated away from food containers and perches. My aviary birds have large stainless steel dishes suspended from the aviary roof by means of three chains. The dishes must have a rim for perching and so that three holes can be drilled to attach a length of chain to each one – most easily achieved with the use of a key ring. Then the three chains can be gathered into a dog lead clip at the top and hung from the roof. These baths are all the more fun because they swing. My lories will even turn on their side and flap around in the water until the two-litre container is empty. In really large pools in zoo exhibits, I have seen lorikeets playing with complete abandon. In fact they look as though they are practising swimming!

Bathing is stimulated by sunshine or rain and is an infectious activity. When one bird starts to bathe the chances are that those in neighbouring aviaries will follow suit. It is good for captive birds because it keep their plumage in prime condition, adds excitement to their day and releases a lot of that pent-up energy.

The degree of enthusiasm that a parrot species has for bathing is partly dependent on its natural habitat. Those from humid climates rejoice in it, while those from arid areas might show little interest. As companion parrots originate from varying climates, the level of humidity in the part of the room where the parrot is kept should be considered. The plumage of a rainforest species, for example, will suffer from being kept close to a radiator. On the other hand, species such as Cockatiels, that live most of their lives in an arid region, are less likely to suffer from lack of water.

Rainbow Lorikeets in a bird park: bathing is an infectious activity!

When I saw an Amazon with dry plumage I asked the owner how often he sprayed it. "About every three weeks", he replied. I pointed out that it needed to be sprayed daily. "I don't have time for that!" he responded. His lack of care annoyed me but I resisted the temptation to say: "Then you don't have time to keep a parrot."

It is sad for a parrot to have brittle, lifeless plumage that clearly shows its environment lacks moisture. Dry plumage and skin can cause irritation and perhaps even lead to feather plucking. Many people take their parrot in the shower with them and, provided that common sense is used, this is an excellent method of supplying much needed moisture.

In the cloud forest

Cloud forests are among the most humid environments. Especially in the Andes, they are home to many parrot species. Calilegua National Park is one of Argentina's three cloud forest parks. Created in 1979, it covers more than 76,000 hectares (293 square miles), protecting peaks and sub-tropical valleys. Tapirs, pumas, otters and more than three hundred bird species live there. It is characterised by a variety of microclimates and, in autumn, I was surprised to find, the highlands are blanketed in fog.

Our vehicle crossed a small stream into the reserve. As we entered the yungas, as this forested region is known, cloud was obscuring everything. In the distance we heard, then saw, a pair of Yellow-collared Macaws *(Primolius auricollis)* in flight. We ascended 1,000m (3,300ft) in two hours, on a narrow road right up to the top. There was silence except for the dripping of water from the trees. You couldn't feel the rain – you could only see it. The cloud gave an atmospheric feel, a mysterious touch to the sopping, moss-laden trees. Here there is up to 2,000mm (79in) of rain per annum and the relative humidity can be as high as 85%.

I was glad to be there, just to see the habitat of the Tucuman Parrot *(Amazona tucumana)* and to revel in the cloud-shrouded beauty of the jungas. Suddenly my guide stopped the vehicle. Five Tucuman Parrots were flying. They soon disappeared into the trees – but I had seen them! This was a special moment for me, to catch a glimpse of what were probably a pair and their three young. It was my sole reason for going there. But I had not expected to find a damp and sunless day like so many of those we experience in Britain.

On the other side of the continent, the Atlantic forest, mainly in Brazil, consists of a wide variety of forest types. Now a World Biosphere Reserve, its area sadly has been reduced from more than one million square kilometres to only 4,000km^2 (1,500 square miles). The annual rainfall varies from 700mm (28in) to 1,600mm (63in). If my experience was typical, it is not a good idea to go there in November! It felt as though the entire annual rainfall, which at the elevation of 800m (2,630ft) averages about 1,400mm (55in) per annum, fell in just a few days.

Parrot watching was virtually impossible. I did glimpse one Blue-winged Parrotlet! This was disappointing with such exciting endemics as the Purple-bellied Parrot in the area. The Atlantic Forest is also home to three endangered Amazon species: the Red-spectacled, the Vinaceous and the Red-browed. The equally endangered and beautiful Red-tailed Amazon *(A.brasiliensis)* lives in a damp world of flooded forest and coastal mangroves.

Amazons are among the parrots that suffer most in captivity if they do not have water on their plumage, preferably several times a week. Those kept in outdoor aviaries in the UK will relish the frequent opportunities to rain bathe. Rainfall in the UK is often generous. The 1981 to 2010 average for Greenwich Park in London was 557mm (22in). The year 2012 was memorable in the UK for the seemingly incessant rain: 1,331mm (52in), the second highest since records began in 1910.

Wind

In 2004 I visited one of the world's oldest rainforests in Malaysia. A very attractive resort can be reached via a 2½ hour boat trip from Kuala Tembeling, near the town of Jerantut, a four-hour drive from Kuala Lumpur. On the right side of the river the forests of the national park stretch for many miles.

It was here that I was able to observe the small Blue-rumped Parrots *(Psittinus cyanurus).* (They are sexually dimorphic in beak and head colour and look something like a *Psittacula* parakeet but with a short tail.) About an hour before dusk they would appear in rapid, direct flight, flying to and fro before choosing their roosting site. One night I witnessed a dramatic scene. Six or more parrots had just gone to roost near a clearing, in a slender tree with sparse leaf cover, when the sky darkened ominously. As I watched to see if they would stay there, a violent wind arose. The tree swayed from side to side, reaching 45° angles. The wind picked up leaves with a rushing sound, until it seemed that it was snowing leaves, then whirled them around and around.

A loud and prolonged cracking sound mingled with the thunder. It signalled the end of a forest giant that had crashed earthwards, making space for another to take its place. Before the cruel wind subsided, the parrots realised they had made a mistake and left the flimsy tree to seek refuge elsewhere.

Parrots hate wind – and this applies equally to aviary birds. The sides of aviaries or blocks of aviaries must be protected. I use corrugated PVC sheeting on wooden frameworks. I put them up expecting to take them down in summer – but they never come down. Britain is an island – so it is windy – and we should never forget that! Most aviary parrots tolerate cold well; their enemies are prolonged damp and wind. Plant trees as windbreaks wherever possible!

Frost and cold

For most people the tropics conjure up an image of palm-fringed Caribbean islands, sun-soaked beaches in Rio or luxury island resorts in the Maldives. These are pictures from the coasts and lowlands but many parrots are highland species where the reality is very different.

Most people know that the Kea is a mountain parrot that lives among snow but they have no idea that dozens of parrot species experience frosty conditions in their natural habitat. In New Guinea, for example, which has a wealth of parrot species, some live permanently above 1,500m (4,900ft). Various lorikeets and other parrots live higher than 3,000m (9,840ft) where frost falls on 30 days of the year, or more.

In Australia, Rainbow Lorikeets live in the Blue Mountains where the winter (June to August) temperature averages 5°C to 12°C (41°F to 54°F) and snow often falls. Gang Gang Cockatoos live in the sub-alpine zone and may descend to a lower altitude in winter. Feeding on berries and eucalyptus seeds aids their survival when snow is on the ground. The ground-feeding cockatoos would not survive such conditions.

In the Andes of South America several parrot species live in the páramo zone. You can look *down* on cloud as I discovered in 2005 in Colombia! This rocky moorland area, where small streams trickle through boggy landscape, is cold – very cold at night. Frost falls. I was with ProAves personnel and we were searching for the nest of one of the world's rarest parrots. This was very easy to find, as we had only to look for a nest-box – and there she was – a female Fuertes's Parrot *(Hapalopsittaca fuertesi),* her little green and blue head and greyish beak emerging from the entrance!

This species had experienced so little contact with man, in its remote habitat, that it knew no fear. In fact, it was believed extinct until its rediscovery just three years before I was there. Despite difficulty of access, most of its habitat has been destroyed. Pro-Aves is turning around the fortunes of this high-altitude parrot which had a population of about 250 birds in 2013. The young and enthusiastic field workers involved in this conservation project are as tough as this little parrot.

The inclement climate is mild compared with that in the habitat of the world's most southerly distributed parrot: the Austral Conure *(Enicognathus ferrugineus).* It occurs as far south as the tip of the continent: Tierra del Fuego. Here it lives in the subantarctic forests of beech *(Nothofagus)* – the nearest forest to the Antarctic itself. There is no clue when you look at this green parakeet that it knows such an extreme climate, with bitter cold and winds up to 100mph. Local people say that when Cachañas fly down from the high Andes, a snow storm is coming. It is true.

I will never forget seeing these parakeets much further north, where the climate is kinder, in the Lanín National Park. I saw them against the glistening snow-covered volcano and the brilliant cloudless blue of the sky, where mauve and pink lupins crowd the landscape like little sentinels.

Lanín is the centre of an area of lakes that stretches for 322km (200 miles) from north to south, on either side of the Argentina/Chile border. An area of outstanding natural beauty, its lakes and mountains are often likened to Switzerland. I never saw parrots in a more picturesque setting! This is one place to which I have always yearned to return.

This ancient forest, where the trees are draped in mosses and the trunks covered in lichen, is made up of evergreen beech, pellin oak and the rich dark

Austral Conure feeding on* Araucaria araucana *(monkey-puzzle tree).

Photograph © Soledad Diaz

green many-forked branches of the Araucaria or monkey-puzzle tree. Here the Cachañas, as the parakeets are known, feed on the seeds in the cones. Alas, all is not well in this idyllic setting. The *Araucaria* trees survive in small numbers and the parakeets were threatened by trapping.

However, alerted by the work of Soledad Diaz and her colleagues, wood extraction regulations in Lanín National Park have been changed to move timber extraction away from nesting areas and to emphasise the importance of holes that might make suitable nest sites. In addition, the continuing education programme in nearby San Martin and Bariloche have increased the awareness of local people regarding conservation of the forests. The work of just a few dedicated people really can change mindsets!

24. ENVIRONMENTAL ENRICHMENT

Environmental enrichment could be defined as any change within a captive animal's environment which is interesting and/or pleasurable. This is especially the case if it goes some way towards encouraging natural behaviours and copying habitat or is conducive to the animal's comfort and physical and/or mental stimulation. This should not be confused with environmental *requirements* which meet the basic needs of a bird and prevent stress or suffering.

The importance of toys

The absence of a stimulating environment not only retards the development of young parrots, it affects their behaviour in the future. They will be less adventurous and inquisitive and more likely to be excessively nervous at the appearance of unknown objects.

Play in parrots has already been described (see **17. Play**) – but the objects used, such as twigs, are limited. Parrots also wrestle and tumble with siblings and other flock members, in actions which seem to have no purpose other than interacting enjoyably with companions. In Australia, cockatoos, such as the Little Corella, which descend to feed, delight in wrestling and rolling on the ground with another cockatoo.

In 1964, Marion Diamond and her colleagues published an interesting paper about brain growth in rats. The neuroscientists had conducted a landmark experiment, raising some rats in boring, solitary confinement and others in interesting, enriched, toy-filled colonies. When researchers examined the rats' brains, they found that those of the "enriched" rats had thicker cerebral cortices. Subsequent examination confirmed that rats raised in stimulating environments had bigger brains. They were also able to negotiate mazes more quickly.

It seem likely that the provision of toys for parrots not only adds to their well-being by preventing boredom, but almost certainly increases their brain power. Parrots should be provided with challenging objects, such as foraging toys whose function is not obvious, and needs to be worked out and remembered.

Other types of enrichment

The value and efficacy of environmental enrichment for Grey Parrots was the subject of various trials by veterinarian Yvonne van Zeeland (see **8. Feather Plucking** and **34. Foraging and Food**). One trial was extremely interesting in revealing what some captive Grey Parrots find most important in their lives. The study was designed to determine how much effort parrots were willing to invest to gain access to specific types of enrichment, as an indication of their relative value to the parrot.

Six Grey Parrots were housed in a two-chamber set-up in which they were able to access specific resources by pushing a weighted door. The effort needed to gain access to the resource was gradually increased until a breakpoint was reached at which the parrot ceased to make successful attempts. Incidentally, the average maximum pushing capacity of these six parrots was 732g (about 150% of the body weight).

There were ten enrichment categories: destructible toys; non-destructible toys; foraging opportunity; auditory stimulation (a radio playing easy-listening music and three sound-producing toys); three cage mates; ladder, rope and net; a large room, 6.5m long; a large bath; hiding opportunity; an empty cage.

Results for three Greys were available. The most effort was expended to get into the large room. However, more time was spent moving around there than flying. Yvonne van Zeeland told me that she thinks this emphasises that parrots need to be let out of the cage to be able to move around freely rather than being in a confined space. Living trees, play stands and the opportunity to go outside would therefore be beneficial.

A lot of effort was put into spending time with other Greys, more or less equalled by food availability. There was no interest in hiding opportunities and only one bird wanted to bathe.

In aviaries

Those who maintain birds in aviaries will receive much more satisfaction from keeping them if they know without doubt that they are doing everything possible regarding environmental enrichment. My heart sinks when I see an aviary containing nothing but a perch at each end. My own aviaries are made into interesting environments. Some contain tough shrubs such as elder growing through the gravel on the floor, or in containers. There are stainless steel dishes hung from the roof to provide not only a satisfying bath but also a swing, and a simple swing made from an apple branch with a length of chain stapled at each end. Depending on how playful the pair of birds is, other items on which they can swing or gnaw are also provided.

Fruits are attached to branches or to stainless steel fruit holders. Throwing fruits or berries on the aviary roof encourages more natural foraging behaviour. I firmly believe that foraging, especially on food items gathered from hedgerows and waste ground, is of immense importance for the welfare of captive parrots and that its importance is nearly always under-estimated or ignored. (See **34. Foraging and Food.**)

By far the best form of environmental enrichment in my own aviaries was a happy accident. A fig seed, from the figs my birds relish, dropped into a crack in the concrete by one of two outdoor cages (with attached indoor cage). This seed grew into a magnificent fig tree that filled one of the cages and grew protectively around both, hiding them from view. By July the fig trees branches (sadly without fruit) are always at their most verdant and extensive. Lucky the two young conures who fledged into this tree!

Other birds as enrichment

Companions of their own species can be considered as a very important form of environmental enrichment. For parrot species that occur in flocks in the wild, a really large aviary (for which few private aviculturists have sufficient space) is the nearest we can get to imitating nature from the important social aspect. Parrots in a flock are never bored! Breeding is usually successful if each pair has enough territory around its nest-box. But the keys to success are a very spacious aviary and the removal of any disruptive individuals.

Unfortunately, due to the aggressive nature of most parrot species, few are suitable. However, many zoos now have Rainbow Lorikeet exhibits, where the public enters and feeds the lorikeets on nectar. These are non-breeding aviaries (there are usually off-exhibit breeding pairs) because most parrots are too aggressive when rearing young (see **18. Territoriality**). For breeders who want the enjoyment of keeping a flock, Budgerigars, Cockatiels, Patagonian Conures, Grey Parrots and some lovebirds are suitable, allowing the birds to interact in the most natural way and to chose their own partners.

The sounds around

In Spain, Xavier Viader maintains a large Grey Parrot breeding facility for the Psittacus company, which produces extruded foods. Everything is thought through scientifically, including the acoustics of the building. To minimise the noise they create, absorbent panels were installed on the walls and ceiling. He wrote: "... the silence at night in a closed room can in itself be a reason for discomfort and alert for these birds which are used to the continuous palpitation of the forest, where silence is usually the prelude of something bad. Therefore, in order to improve the acoustic comfort, we installed in the bays equipment which plays recordings made in the forests of origin of this species, thus trying to simulate as much as possible the auditory framework that differentiates each particular time of the day and of the night" (Viader, 2010).

This is a novel approach by someone who obviously cares a great deal about the environment in which his birds live.

Music is another form of environmental enrichment to which most parrot respond and vocalise happily. I often have a radio/CD player playing in my birdroom – preferably classical or light music. When the music stops the birds instantly stop vocalising. As Xavier Viader said: silence is usually the prelude to something bad.

25. THE WAKING HOURS

Parrots are birds of the tropics and sub-tropics. They occur, with few exceptions, in a wide band above and below the Equator from about 30° north to 30° south. This means that most species experience twelve or thirteen hours of daylight. Very soon after sunrise – or even before – they start to leave their roosting sites.

The Critically Endangered Puerto Rican Parrot has been intensively studied since the late 1960s because it came so close to extinction. Observations of one small group of about eleven birds in the Luquillo Forest in 1971 revealed that they started to become active just before 7am; sunrise was at 6.52. Their afternoon feed commenced at 15.45 and continued by at least one pair until 17.25. Soon after, on this overcast day, all the parrots had returned to their roosting sites (Snyder, Wiley and Kepler, 1987). This gave a day length of about 10½ hours but foraging could commence earlier to give a day length of 11 hours.

The latitude of Puerto Rico is 18°N. Compare this with between 50° and 59°N degrees for the UK. Patagonia, in southern Argentina, has latitudes between 37° and 51°S with daylight lengths for winter and summer almost the same as those in Northern Europe. Patagonian and Chilean Conures live here. As already mentioned (**23. The Elements**), the Austral Conure has the most southerly distribution of any parrot, stretching as far as Tierra del Fuego. Daylight lasts for seventeen hours in the summer and only seven hours in the winter.

The daylight lengths are therefore completely different to those experienced by parrots of the tropics although the temperatures in the tropics vary, of course, according to the altitude – and many parrots are montane species.

Day length in species also varies according to their habits. When I was in my twenties I kept Bourke's Parakeets, which I found delightful and I loved their "exquisite colouring". I kept notes about all my birds. I wrote: "They love the twilight and their big black eyes provide a clue to this." I would hear them twittering, in their appealing way, at nearly 10pm on summer evenings. This is unusual in that most aviary birds seek their roosting site well before the light fades.

These parakeets are crepuscular with eyes adapted for this habit (see **5. The Eyes have it!**) so their day length differs. But do truly nocturnal parrots exist? The answer is yes – there are two species but sadly both are struggling to survive in a world vastly changed by man.

Nocturnal parrots

One species made sensational news among ornithologists in 2013. The Night Parrot *(Pezoporus occidentalis)* is ground-living, 23cm, long-tailed – and one of the most secretive birds in the world. Discovered in 1845 but rarely seen alive, it inhabits the arid interior of Australia. It had not been reliably reported for decades when optimism regarding its survival was sparked with the discovery of a desiccated corpse in 1990 in

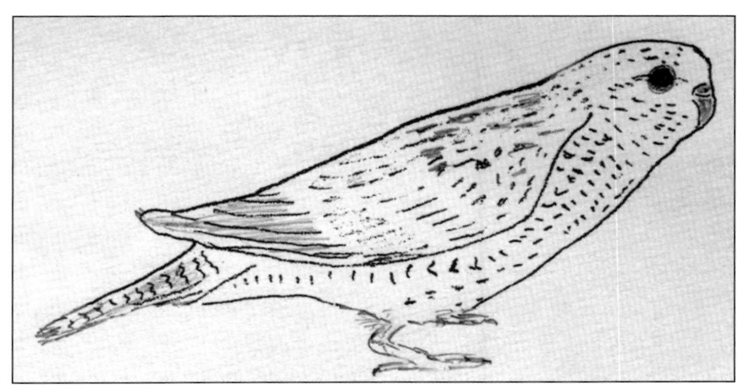

Night Parrot.

remote north-western Queensland. An awareness campaign was launched to encourage travellers to look out for it and a big reward was offered for proof of the bird's survival.

On July 3 2013 moustachioed John Young presented photographs and six seconds of video footage of Night Parrots to an amazed, selected audience of scientists. This happened at the Queensland Museum. Young had spent, he said, 17,000 hours and 15 years in the field searching, finally studying it at an undisclosed site in western Queensland – undisclosed because of the danger to the birds and the fragile environment if it was invaded by people. Young discovered that it truly is a night parrot – not emerging at dawn and dusk as was previously thought.

Controversy had surrounded one of John Young's bird sightings in the past but "This time", he said, "I took the memory card out of the camera and locked it in the bank! This time I have undeniable proof." He also made recordings of its vocalisations.

Sean Dooley, editor of *Australian Birdlife* magazine, told the Australian Broadcasting Company that the discovery (if proved true) was "the equivalent of finding Elvis flipping burgers in an outback roadhouse."

The other nocturnal parrot is New Zealand's Kakapo *(Strigops habroptilus)* – arguably the strangest and most endearing parrot that ever lived. Technically extinct in the wild, it exists only due to intense management on islands cleared of predators. It, too, is a truly nocturnal species, hiding away in thick vegetation during the day. At any one time, its population is known precisely – 127 at the time of writing (July 2014) – but that is another story.

Structure to the day

Parrots and other birds have an extremely well developed sense of time. The structure of the day of most parrot species never varies: leave roost site to feed; rest, preen, bathe and socialise during the middle of the day; feed again and find water during the late afternoon, then fly to roost.

I believe that our companion parrots should also know a routine: that they should be fed at the same times every day. Their sense of time is truly remarkable and they will become impatient if the feeder is late!

For aviary birds without access to accommodation with lighting the day length will relate to the hours of daylight. For those who reside with humans, or who are shut in their well-lit inside quarters at night, the hours of light, artificial or natural, should not fluctuate too much.

It is not natural for the majority of species to know more than about thirteen hours of daylight. In companion birds the parrot's cage needs to be covered or a sleeping cage in a room away from bright lights, television and other high levels of noise and light should be used. Tiredness can be caused by excessive light, especially in young birds. In parent-reared young of species that roost in the nest-box, recently fledged young will retire up to two hours earlier than adults. They don't need to be told to go to bed early!

According to the latitude from which parrots originate, sunset can last only a few minutes or it can be gradual or prolonged. It is never instant! Most birdkeepers realise that suddenly plunging birds into darkness can have serious consequences, causing them to panic and perhaps injure themselves. In parrots kept inside buildings during breeding operations eggs or chicks could be deserted. In northern climes birdroom dimmers are invaluable during the winter, allowing the lights to dim over a set period. This gives the occupants plenty of time to settle down for the night and, in the case of small birds who might find it more difficult to maintain their body temperature overnight, to have a last feed.

PART V.
BREEDING

A female Budgerigar helps a chick out of the egg.

26. LIFETIME MONOGAMY

Many – in fact most – bird species are monogamous. This means that they bond with another bird and remain with it for the duration of that breeding season. Lifetime monogamy is much rarer, as is genetic monogamy – where all young are sired by one male. This is unusual in birds and occurs predominantly in some non-passerine groups which have long reproductive life spans and in which males also care for the young This applies to most parrots. One advantage of lifetime monogamy is that if, for example, due to weather conditions which will soon result in a very good growth of rearing foods, pairs that are permanently together can start to breed immediately. Such pairs often maintain their nest in good condition throughout the year.

In lifetime monogamous parrots mutual preening is an extremely important part of maintaining the pair bond. Male and female usually perch in close body contact and preen each other's feathers for minutes at a time.

In some bird species up to 50% of the young are not fathered by the male of a bonded pair. It is often said that "parrots pair for life" – but is this true? Until fairly recently, there was no way of marking these birds and following them over time. Now, thanks to the research of some scientists who devote most of their working life to one particular species, we are beginning to have some answers.

In 2003 I spent a few memorable days in Patagonia, at the famous breeding colony of Burrowing Parrots (Patagonian Conures) at El Cóndor. I was able to observe the researchers at work – Juan Masello, an Argentine biologist, and his German wife Petra. They had been studying this colony, which they described as the largest parrot colony in the world, since 1998. Remarkable for their enthusiasm and endurance, their on-going research is ground-breaking in many aspects.

It seems that they can go some way to answering the question of lifetime monogamy in respect of these large, colourful and elegant parakeets. Some of their nests in the sandstone coastal cliffs, facing the Atlantic Ocean, are accessible. These have been numbered and the chicks monitored until they fledge. It is possible to handle them, thus DNA finger-printing had been carried out on forty-nine families by 2002. A total of 166 chicks were sampled. There was no doubt that 151 chicks belonged to the putative parents.

Novel fragments in the DNA of thirteen were believed to be the result of mutation. However, two of the chicks were not related to the birds who reared them, suggesting that a female had laid an egg in the nest of another pair. In one case, there was a connecting chamber that led to a neighbouring nest. Collapse of the wall between two nest chambers could also produce this result (Masello *et al,* 2002). The conclusion was that no extra-pair paternity, as scientists call it, had occurred. So they are faithful to their partners!

This still does not answer the question of whether parrots pair for life. Some years ago American scientist Steve Beissinger was studying Snail Kites in Venezuela when he saw parrotlets nesting in a fence post (see **15. Communication**). It occurred to him that if he provided nest sites, these birds would be easy to study. This fortuitous observation resulted in a twenty-year research programme that revealed amazing facts about these sparrow-sized parrots.

I was fortunate to be present at the 2007 Parrots International convention in Long Beach, California, when Steve Beissinger presented some of his findings. By then he and his researchers had ringed 7,500 individual Green-rumped Parrotlets and tracked 3,000 nest attempts. (Surely a feat unequalled in any other parrot research project.)

It seems that parrotlet pairs are very faithful. Of the 488 pairs studied from 1988 to 2003, 83% of pairs stayed together. In 11% the female found a new mate if the male died but only 5% of males were able to find a new mate in that event, due to an excess of males in the population. Amazingly, "divorce" occurred in only 1% of cases. This is an extremely interesting statistic resulting from research that would be impossible to conduct in most wild parrots.

However, in his 2008 paper, Beissinger wrote that 75% of 738 marked pairs nested together for only one year due to the death of one of the pair. The annual survival rate was only 62% (Beissinger, 2008). This is probably the only study that has recorded annual mortality in a parrot population. I suspect that in larger parrots, with their smaller clutch sizes (average seven in the parrotlet) and, in some species, not breeding every year, annual mortality would be less. The parrotlets usually had two successful clutches per year.

Research on the Western Long-billed Corella *(Cacatua pastinator)* showed a different story – but it was limited to thirty-nine pairs. "Divorce" occurred in six pars (15.4%) during the period 1977 to 1982, while in another five pairs one partner disappeared, and presumably had died. A "divorce" rate of 25% was recorded for 16 pairs in which both birds were tagged and were known to have survived from the first year they were together to the following year. In all instances the pairs split up following the first breeding season, during which three pairs successfully reared young (Smith, 1991).

Does the term lifetime monogamy mean that male and female stay together throughout the year or are they nomadic, just getting back together for the breeding season? In almost all the parrot species that have been studied, male and female do exist side by side. An interesting exception is the big New Zealand forest parrot, the Kaka. A study at the Waihaha Ecological Area showed that male and female move about independently for much of the year, from March to August. The pair bond usually remains stable during the breeding season, until March, when any young become independent (Greene and Fraser, 1998).

But how about those in captivity? What does the breeder need to know? Because captive parrots cannot choose their mates, there are many incompatible pairs. Ideally, a number of parrots would be placed in one aviary and allowed to form pairs. Unfortunately, this is rarely possible. Some breeders allow parrots to fly in large aviaries outside of the breeding season when it has been known for them to change partners. My guess is that this would not happen with compatible pairs.

If one bird of a long established compatible pair dies, it can be very hard to find a new partner that would be accepted by the widowed bird. So, yes, in this case I believe that many captive parrots do pair for life or they just tolerate a new partner, rather than form a strong bond with it.

27. MATRIARCHAL SOCIETIES

The temperament of individual parrots varies, just as it does in humans. But there are some elements of behaviour that are fixed and peculiar to certain species. Unlike most parrots, Eclectus belong to a matriarchal society. They are remarkable for the dimorphism of their plumage: the male is green and the female is red. Males are popular as pets and have a different temperament to females. Birds of both genders can learn to talk and imitate. However, there is a problem with females as companion birds – and not just because they can be a little bad-tempered! When mature, they would like to spend all their time under the newspaper on the cage floor or hiding in any small space.

To know the reason we have to understand their breeding biology, which was poorly known until the early years of this century. Native to New Guinea, some Pacific islands and Australia, Eclectus need hollows in large emergent trees for breeding purposes. Such hollows are not common. Researchers in Cape York, Australia, surveyed Iron Range National Park from a light aircraft and found a maximum of only one suitable nest tree per square kilometre of rainforest. When a female finds one and starts to use it (probably having usurped another female) she will rarely fly more than a few metres from that tree, because if she leaves it unattended another female will try to take it over. The attempt could lead to a fight to the death. Females are therefore more aggressive and assertive than males.

How can a female live like this? When does she feed? She does not forage for herself but is totally dependent on several males (four or five) to feed her. One male alone would not be able to provide

sufficient food for her and her two chicks. For many months she will be tied to the hollow, producing more eggs and chicks until she is evicted or until the tree hollow floods.

Why has such extreme plumage dimorphism evolved in Eclectus? This is almost certainly because males spend a large part of their waking hours foraging and green plumage probably is the most effective camouflage. Adult females spend much of their lives in a nest hollow so camouflage is not important. But when they leave that hollow they need to be conspicuous so that they will be fed promptly by a male.

In Eclectus the female is the central being in the group. There is no strong pair bond, and generally a lack of the affectionate behaviour towards a partner seen in most other parrots. This is also true of *Psittacula* parakeets, such as the Ringneck, and a few other parrots.

This has significant implications when keeping these species. For breeders it means that because the male tends to be wary of the female, he should be placed with her well before the start of the breeding season. It takes a while for him to become confident enough to start his ritualised display behaviour. For the breeder it also means that pairs do not have to be maintained permanently, but the male can be placed with a different female in any year, usually without the stress involved in breaking up a bonded pair of other parrot species. In Eclectus, a female will breed in a large aviary with several males. This would be the most natural method – not one pair per aviary in which the male can become stressed by the female's overbearing behaviour.

For the companion parrot owner the difference means that males usually make better pets and females can be bad-tempered at times. They can be sexed when the contour feathers start to appear at about four weeks.

A female is likely to start laying as soon as she finds a cavity. She will not want to come out. If she lays, her two eggs should not be removed as this will only result in her producing more. There could be long periods when she is not prepared to socialise with her human companion. However, there are many cases of female Eclectus and lovebirds (which show these traits to a degree) bonding quite strongly with the owner and making lovable companions.

In this group of parrots – Eclectus and *Psittacula* parakeets – mutual preening does not usually occur, so they do not like to have the head scratched by their human companions – something which affectionate species crave. They do not like to be handled too much – which, actually can be an advantage for some people, especially as it means they are less demanding than Greys, Amazons, macaws, etc.

From Madagascar come the strange grey Vasas *(Coracopsis)* – the least colourful parrots in existence. They are also, behaviourally, the most bizarre – next to the Kakapo. When they were in my care I found them so endearing and interesting! Their love of sunbathing and the strange positions they assume when engaged in this activity (see page 109), and their unique breeding biology, easily compensated for their lack of colour and unusual appearance.

Uniquely among parrots the male has a phallus, visible only during copulation. It is an extraordinary sight – rarely photographed – so that I could not lose an opportunity to record and show this. It could be mistaken for a prolapse and harmful and unnecessary action might be taken. The other unique fact in the world of parrots is that male and female remain joined together for a very long time – more than one hour in some cases.

The first time a pair of Lesser Vasa Parrots *(C.nigra)* produced young at Loro Parque, Tenerife, the female sadly died when the chick was a few days old. At that time it was not known that the female is attended at the nest by several males. I believe this female's death was due to the fact that she did not leave the nest to feed and the single male did not provide her with sufficient food. In captivity it is therefore of the utmost importance to place more than one male with the female.

It is interesting that this species has the shortest incubation period of any parrot – no more than 14 days. The young spend a shorter time in the nest – only four or five weeks. They grow at an astonishing rate. Little is known about their breeding biology but the short incubation and rearing periods suggest that they might feed on a seasonal food item which is available for a very short period, thus the rearing period must be brief. In order to achieve this, the female needs to be fed by several males. Or perhaps the rearing foods are not abundant or they are widely dispersed.

Phallus of a male Lesser Vasa Parrot. This is not a prolapse.

Female dominance

In most parrot species the male is usually the dominant member of the pair. As mentioned above, this is not the case with *Psittacula* parakeets, or with lovebirds. In addition, in most *Poicephalus* parrots females are dominant and bolder. Keeping two females together must be avoided. I learned this the hard way with two female Red-bellied Parrots which were placed in one cage for quarantine. One soon killed the other. There is no problem in keeping two males together. Indeed I can recall a breeding pair of Grey-necked Parrots at Vogelpark Walsrode, Germany, in the 1970s in which a second male was kept in the same enclosure as the breeding pair.

Comparatively few parrot species have been studied in the wild. At this point in time there are many whose breeding biology is hardly known. Could it be that Eclectus and Vasas are not the only parrots whose females are attended by more than one male? As most parrot species are not sexually dimorphic, and many nest high in the canopy, this might not become apparent until intensive studies of little-known species are made.

28. MALE PARTNERSHIPS

Steven Beissinger's research on the Green-rumped Parrotlet *(Forpus passerinus)* commenced in 1998 on a scale never before or since achieved. In the following nineteen years he marked and re-sighted 7,500 individuals. His study highlighted the excess of males in the population. "Pairs" consisting of two males apparently comprised 10% of the non-breeding males that were repeatedly seen. Typically they consisted of young birds under the age of one year (79%) or one to two years (16%). Only 6% were composed of known siblings.

Male-male pairs were seen together for up to one year but most of these friendships lasted only two months. Male pairs parted when one male obtained a female (63%) or when one died (19%). The other reason for the end of these partnerships (12%) was when one male befriended a new male. Non-breeding males were difficult to detect and to follow (Beissinger, 2008).

In captivity there are more males of most species. This means that two males are often kept together, either in the mistaken belief that it is a true pair, in a group or for other reasons. In species which have a strong pair bond, such as most neotropical parrots, two birds which often sit close together and preen each other and perhaps even copulate, could be two males. If the bond lasts a long time it might be difficult to introduce one to a female, but more often than not a breeding pair can be formed. Some people refer to two males with a strong bond as a homosexual pair but this is seldom true as in different circumstances the male would have formed an equally strong bond with a female.

Today most breeders do not know how lucky they are! Using several plucked feathers or a small blood sample, DNA technology is used to quickly and easily identify the sexes of parrots, via a laboratory. When the revolutionary but invasive surgical sexing technique became available in the 1970s, many breeders found out why some birds had never produced eggs! Two males had been masquerading as a "pair". The problem has not gone away entirely because some unscrupulous breeders or dealers sell two males as a true pair. Wise purchasers ask for DNA sexing certificates with ring numbers that relate to the birds in question.

Same sex preferred in human companion?

Do parrots prefer the same sex or the opposite sex in their best human friend? A companion bird in a family often forms a very strong attachment to one person, to the exclusion of everyone else. I suspect that opposite sex attachments are more common – but so many factors influence this. Sometimes a parrot seems to make an instant decision that one person will be his or her favourite. This could be influenced by past experiences, the appearance of the person (reminiscent of someone else?), whether they are empathetic (parrots are quick to discern) and the person's voice.

It is a great disappointment for many people that the parrot they bought prefers their partner!

29. NEST SITES

Most parrots nest in holes in trees: cavities formed when branches break off and woodpecker holes. The entrance often determines the choice of nest: if it is too large, it will be ignored. If it is too small, an attempt might be made to make it larger. Female parrots may take a long time to prepare their nest site. They remove unwanted material – perhaps from a previous nesting – and might enlarge the cavity or the entrance.

Selective felling of mature trees in the tropics often makes it difficult for large parrots to find a suitable nest. This can result in breeding failure if a nest with a large entrance, and therefore more susceptible to flooding and predation, is chosen. It can also result in parrots nesting in closer proximity than would be normal.

In Costa Rica, where so much lowland forest has been felled in recent years, an *almendro* (almond) tree *(Dipteryx panamensis)*, was the subject of much interest in 2011. Isolated and 100m (328ft) from the forest edge in a clearing, it contained *three* active macaw nests – probably a unique discovery. Two were inhabited by the Endangered and magnificent Great Green Macaw, only a few metres apart. In the third a pair of Scarlet Macaws were rearing young, lower down in a major branch. The tree was in a state of deterioration with many cavities, some of which contained nests of Crimson-fronted Conures *(Aratinga finschi)*.

The *almendro* is the main nesting site of the Great Green Macaw and two nests in the same tree occasionally occurs. When one considers that the total population of this macaw in Costa Rica is about 300 individuals, it gives an indication of the scarcity of nest sites.

The three-nest tree was observed by Olivier Chassot and Guisselle Monge-Arias, whose field work for more than a decade, commencing in the 1990s, was instrumental in protecting what was left of the Great Green Macaw's habitat in Costa Rica. In 2002, I went with this dedicated couple to the area in which the tree described above is located. I was deeply saddened to see huge felled *almendro* trees rotting in paddocks where they had been cut, although it is illegal to fell them. Nest trees have inscribed plates attached to them, in the hope that they will be protected.

I was fortunate to be there (in the torrential rain), when a young macaw looked out of a nest spout. It was a thrilling moment for me and one of deep satisfaction for Olivier and Guisselle, knowing that the population was about to increase by two. The first of the youngsters fledged the next day (Low, 2009).

Hazards at the nest

In Ecuador the Great Green Macaw survives in extremely small numbers. In 1993 a programme to conserve it was launched by Fondacion Pro-Bosque. In the Cerro Blanco Protected Forest a pair nested in a dying pigio tree *(Cavanillesia platanifolia)* at a height of about 20m (65ft) from the ground. At the end of October a young one was seen looking out of the nest and the female then began to leave it for extended periods.

This was risky because during the rearing period a pair of Grey Hawks nesting not far away had attacked the male. In addition from mid-July a pair of Collared Forest Falcons were seen near the nest. The larger female falcon entered the cavity on August 6 and was driven off by the male macaw. That month falcon attacks became frequent – they dived on the macaws and struck them with their beaks. They even grappled with the big parrots. By November 6 the falcons were preventing them from approaching their nest and would sometimes pursue them more than 180m (600ft) from the nest.

When you consider that the macaw, at about 1,300g, is the third heaviest parrot with the second largest beak, it could be questioned why the macaws were bullied by this falcon. It is almost exactly half their weight and half their length 46-56cm (18-22in). One can only assume that the fearsome talons of the falcons were the reason.

The young macaw was desperate for food during the two days of the falcons' siege and was attacked at the nest entrance. On November 8 it was dragged out of the nest which was taken over by the falcons. A researcher observing the macaws from a nearby hide rushed to save the young macaw as it was attacked at the base of the nest tree. It was unharmed and then spent two days in the house of Eric Horstman (director of Fundacion Pro-Bosque) where it was fed and cared for. Two days later it was placed in the nest area when the parents were nearby. Very soon the young macaw flew off with them so there was a happy ending after all (Horstman, 2014 and pers. comm. 2014).

This amazing incident illustrates some of the perils and difficulties faced by parrots in trying to care for their young.

Nest builders

The Quaker or Monk Parakeet is well-known for its habit of actually constructing a nest without the existence of a hollow. Its bulky stick homes have brought down power lines, plunging whole cities into darkness. It is fascinating to watch aviary birds cutting twigs and weaving them into position.

Some lovebirds (the white eye-ring species) actually build a nest inside a cavity, nipping off grasses and carrying them back tucked into the feathers of the rump. The result is a neat domed nest. In captivity breeders provide willow twigs for this purpose.

Unlike all other parrots, the Palm Cockatoo builds a platform of sticks as a base for its open-topped nest. It fashions twigs of the correct length and drops them inside. The resulting platform which, in wild birds varies from a few inches up to 3m (10ft) in depth (according to the height of the nest hole), protects the single egg or the chick from flooding. In captivity this species is unlikely to breed without the stimulus of a daily supply of fresh-cut sticks during the appropriate period. A pair in my care at Palmitos Park took this habit to extremes. They filled the nest log to within 15cm (6in) of the top, which could have resulted in the chick fledging too early. Sticks were therefore removed to give a 31cm (12in) deep nesting area – and eventually a chick fledged safely.

Nest excavators

Few parrots actually excavate nest sites in trees. Among those that do are fig parrots; they are small but they have powerful beaks. In contrast the large red and black Pesquet's Parrot *(Psittrichas fulgidus)* from New Guinea excavates a new nest site every year in a dead tree trunk. I have been lucky enough to work with these totally wonderful parrots (my favourite species!) in two locations. The stimulus for breeding was the provision of a palm trunk with an indentation in the side.

The pair that I cared for at Palmitos Park, Gran Canaria, were provided with a solid palm log 2.4m (8ft) high. The female was seen at the nest for the first time on December 17. Within a few days both birds were spending long periods inside, excavating. Every day the aviary floor was littered with soft pieces of palm fibre. By January 1 they had excavated to a depth of 50cm (20in), by January 5 to 65cm (26in) and by the 7th to 75cm (30in). I became concerned about the female who was coughing frequently, probably because of fine fibres ingested during excavation. Wood shavings placed in the nest and dampened with a bucket of water solved the problem.

By January 19 the pair had excavated to a depth of 1.1m (44in) and all the wood shavings had been removed. Male and female would shake their feathers and a cloud of fibres would fall to the ground. The excavation was 1.25m deep by January 24 and 1.4m by February 1 when they had almost reached the bottom. The finished interior

Smaller nest log excavated by the Pesquet's Parrots after their initial success (right). Below: the first Pesquet's reared at Palmitos Park, aged 13 weeks. The author's favourite bird!

surface was neat and even – almost as though it had been machine-turned! The female laid her first egg on February 4 (Low, 1992). The pair went on to produce several young.

I have described the nest preparation in detail because it is such a rare event in the world of parrots. It is remarkable also because the beak of Pesquet's Parrot does not look powerful like that of a macaw or a cockatoo: both mandibles are unusually elongated.

Very little research has been carried out on the breeding biology of parrots in Papua New Guinea. Notable is the work of the late Paul Igag, a Papuan who studied conservation biology at Crater Mountain, from 1999 for his Master's degree (supported by the Wildlife Conservation Society, New York.) By training twenty-three traditional land owners as research assistants he was able to monitor the nests of 71 Eclectus Parrots, 51 Palm Cockatoos and 34 Pesquet's Parrots. The work continued during 28 months on a study site of 2,645 hectares (1,058 acres). It highlighted the scarcity of available nest sites (the table below), also the fact that Pesquet's Parrots prepare a new nest every year.

per hectare	**density of nest trees**	**annual density**
Palm Cockatoo	0.06	0.008
Eclectus Parrot	0.069	0.023
Pesquet's Parrot	0.017	(new nest annually)
Young fledged as % of eggs laid		
Palm Cockatoo	40%	
Eclectus	54%	
Pesquet's	17%	
	(results for only 20% of broods)	

Nest entrance size

Contrast this study with the lifetime interest of a single individual. In New South Wales John Courtney, in his late seventies when this book was published, has spent sixty years studying nesting Little Lorikeets *(Glossopsitta pusilla)* and Musk Lorikeets *(G. concinna)* on what was formerly his farm in Swan Vale, Glen Innes. Here I should pay tribute to his work which, for continuity of the study of one parrot species, entirely unaided, is surely not surpassed by any other person on the globe.

Locating a nest of this species is notoriously difficult. As I related in *Encyclopedia of the Lories,* over the years he measured the entrances of countless nests on and near his property. Up to 1989 he had found nearly fifty breeding cavities of the Little Lorikeet. In 1980 he told me:

"... the difference in diameter between the nest entrance of the musk and little lorikeets is about 9mm (3/8in). This does not seem very much, but is probably of crucial importance in preventing musks from taking over little lorikeet nests. I have noticed that both species have to wriggle a bit to enter or leave their nests" (Low,1998). Using the minimum size nest entrance helps to deter predators. In one Musk Lorikeet nest, the entrance measured 4cm (1 5/8in) vertically – just wide enough to admit a Musk's shoulders, but only if it turned side-on.

Nesting within plants

By far the majority of parrots nest in holes in the trunk or branches of trees. They are easy to cater for in captivity because they readily accept nest-

Little Lorikeets at their nest at Swan Vale, New South Wales. Photograph: John Courtney

boxes. The same is true of certain species that are prepared to use less conventional nest sites. They are adaptable, prove very prolific in captivity and some (but not many) can now be described as domesticated. This includes the Eastern Rosella – often called Golden-mantled Rosella by aviculturists. Unusual nest sites of this Australian species include rabbit burrows, nest tunnels of bee-eaters, an arboreal termitarium, cavities in buildings and crevices in rock faces in a quarry. The epiphytic elkhorn or stagshorn fern was used as a nest by one pair! It was located on the wall of a house.

In Sulawesi, Indonesia, a nest of the Crimson-spotted (sometimes called Yellowish-breasted) Racket-tailed Parrot *(Prioniturus flavicans)* was discovered in December 1997. Situated in a cavity in the root ball of an arboreal epiphytic fern at a height of 28m (92ft), it was growing on a strangler fig tree. The fern's root ball had a diameter of about one metre (33in) with a circular cavity entrance in its underside, about 12cm (5in) in width. Three young fledged from this nest, between February 24 and March 3. This was probably not an unusual site for this species as at three other locations pairs were seen exploring similar cavities (Walker and Seroji, 2000).

Racket-tails are not the only parrots to use root balls in epiphytic ferns. The related *Geoffroyus* parrots and the unrelated *Charmosyna* lorikeets also do so. Not many zoos or aviculturists would have the opportunity to provide such a nest site. However, at San Diego Zoo in California I was interested to see that a pair of small lories were nesting in a box that was entered via a stagshorn fern!

In Queensland Pale-headed Rosellas *(Platycercus adscitus)* used a similar site, the back of the cavity being formed by a section of tree-fern trunk on which the elkhorn was mounted. The brown tips of the dying fronds curled back over the cavity, providing security and shade (Carter, 1996).

The use of termitaria

Some parrots with unusual but precise requirements are more difficult to breed in captivity. Throughout the tropics, many birds nest in termite mounds. This includes several species of the little *Brotogeris* parakeets. In Peru, Tui Parakeets *(B. sanctithomae)* and Cobalt-winged Parakeets *(B. cyanoptera)* usually use termitaria containing live termites because abandoned nests are liable to crumble and fall. In the Amazonian forest of Ecuador, a native Quechua Indian guide told me that Cobalt-wings always nest in termitaria. Most neotropical parakeets breed readily in captivity – unlike these two species. Some *Brotogeris* breeders have found that horizontal nest-boxes, with an inner chamber, are preferred and might be an incentive to reproduce.

In aviculture

In a captive parrot, a sudden or an increased urge to gnaw wood is usually a sign that it is in breeding condition and following its instinct to excavate. It is very frustrating for a parrot to be unable to gnaw wood at this time so fresh-cut branches or off-cuts of untreated timber should be available. Pairs that already have a nest-box might do considerable damage to it unless pieces of timber are screwed on the inside or outside. Beware the use of nails for this purpose as they will be exposed when the wood is shredded.

Parrot breeders should provide nest-boxes with the smallest holes possible. Many parrots, especially non-domesticated species, are deterred from entering boxes with large entrances. I recall the case of a pair of rare, delightful and diminutive Yellow-lored (now also called Yucatan) Amazon Parrots *(Amazona xantholora)* at Palmitos Park. This is the smallest of the genus. They had been provided with a standard Amazon parrot nest-box which they had never entered. I asked the keepers to nail a piece of wood over the side of the entrance, reducing the width to 6cm (just over 2in). The pair entered the box soon after and went on to produce young every year. An entrance which is too large not only permits too much light to enter but goes against a parrot's instinct for the smallest possible opening.

African Lovebirds are ready breeders, so much so that many mutations have been developed. There is one exception – the Red-faced Lovebird *(Agapornis pullarius)*. It breeds in holes (usually excavated by woodpeckers) in ants' or termites' nests in trees, or occasionally in those constructed on the ground. A tunnel leads from the entrance to the lovebirds' nesting chamber.

Breeders have found the best method is to pack a nest-box with cork or some other material in which the lovebirds can burrow. Excavation is probably a great stimulus to breeding. The first success which definitely related to this species occurred in 1955. Arthur Prestwich suspended four bales of peat in the aviary in which the lovebirds burrowed. And one youngster fledged!

In aviaries it often happens that parrots want to nest on the ground. This can happen if no nest-box is provided, causing the pair to tunnel into the earth floor of the aviary. Such floors are not recommended because they are unhygienic, not vermin-proof and can result in escapes. But some bird keepers seem to be sublimely unaware of these risks!

One breeder offered his macaw pairs a choice of three sites – a conventional nest-box, a log and a brick-built nest at ground level. Different pairs chose different sites. Some parrots, especially macaws, readily use boxes at ground level. Years ago my Goffin's Cockatoos refused to enter any nest-boxes so in desperation I placed a natural log, with wooden entrance added, on the floor. They entered, reared young and used it for many years!

Rainbow Lorikeets are among the species that might make a nest in the aviary floor – if the keeper is not vigilant. In the Admiralty Islands of New Guinea Olive-green Lorikeets *(Trichoglossus haematodus flavicans)* nest and roost on the ground on at least three small islands, in rock crevices or under a rock overhang with no nesting material. These risky sites are used because there are no trees large enough to provide cavities.

The readiness of many parrots to use unorthodox nesting sites is one reason why many species breed well in captivity. They are opportunists and will accept various sites if the entrance is the right size.

30. FROM LAYING TO HATCHING

What motivates parrots to breed? They need a compatible mate, a suitable nest site and a reliable food source. If the season when they would normally be breeding is too dry, the whole population will make no attempt to breed because the fruits, flowers or seeds they need to rear their young will not be produced. These items are usually those which have a higher protein content, as protein is essential for the growth of chicks. For most species, missing a breeding season every few years will have no impact on the population. For endangered species which survive in seriously low numbers it can be disastrous.

By 2008 the Critically Endangered Blue-throated Macaw (see illustration on page 6), found only in Bolivia, had a total (world) breeding population of only 15-20 pairs. The subject of intense conservation measures since 1995 (its area of origin was unknown until 1992), this stunningly beautiful macaw had been trapped almost to extinction in the 1980s. Its main food source is the fruits of the motacú palm *(Attalea phalerata)* but 2008 was so dry that these palms produced no flowers or fruits. Only one pair attempted to breed and sadly the single chick had a deformed wing.

Food is not the macaw's only problem. In such a small population scattered over a large area, a young bird might find it difficult to locate a suitable mate.

Lack of nest sites can be addressed – but only to a small degree. Of thirty active nests monitored by researchers over four breeding seasons, 43% were predated (usually during the incubation period). Most artificial nests erected for these macaws as part of the conservation project were occupied by bees or other bird species, but in 2007 one pair used a box and reared *three* young. Other nests were made flood-proof; chicks die when nests become waterlogged. Parrots face many problems when rearing their young. Usually only in endangered species which are intensively studied are these difficulties revealed.

Sexual maturity

The age at which parrots mature sexually is related to their lifespan. A lovebird *(Agapornis)*, Kakariki *(Cyanoramphus)* or small lorikeet, with a potential lifespan of about 20 years, could produce young when only six months old. They have been known to do so in immature plumage. A large macaw (potential lifespan sixty years) would be unlikely to start to breed until aged between four and seven or even eight years. However, cases of female macaws breeding before three years old are not unknown.

To find out when wild parrots commence breeding means studying one population over several years, thus records are sparse. It is known that male Galahs first reproduce aged only two to three years and females a few months later (Rowley, 1990).

Why do captive-bred parrots become sexually mature at an earlier age than is desirable. First of all, how would we define "desirable"? Budgerigars and lovebirds, for example, are sexually mature at the age of six months but responsible parrot breeders do not want their females laying at such a young age. Physically or psychologically they might not be mature enough to breed well, to incubate properly or to have the instinct to feed their chicks. There might be a comparison here with young teenagers in humans, who are sexually mature but not responsible enough for parenthood.

It has been suggested that pet birds might mature early because they are responding to handling and stroking. (They should never be stroked on the back as this could evoke a similar response to copulation.) This theory does not explain why some birds that are not handled also mature at a young age.

So what does stimulate some captive birds to mature very early? In most species it will be the presence of a mate (especially an older one) and a suitable nest site. Most breeders will not supply nest-boxes until they consider the birds mature enough to breed – but for species that roost in their boxes, such as conures, caiques and lories, nest-boxes might always be present.

Certain species, such as white cockatoos, are stimulated to breed when it rains whereas for others the stimulus will be a protein-rich diet. Some pet birds might receive a diet that is too high in protein. In addition, they receive longer hours of light (artificial) than they would in their natural habitat. This is another stimulus for many species.

Mate choice

The Budgerigar is unique in having a cere (the fleshy part immediately above the beak), in which the colour is different in adult males and females. Normally (with the exception of some mutations), in birds in breeding condition the cere is deep brown in a female and bright blue in a male. The dark brown colour indicates that the female's ovaries are well developed, as Budgerigar breeders know. Beginners might be puzzled by a female who has a light brown or even a light blue cere. This indicates that her ovaries are undeveloped. Most male Budgerigars also know this. In an experiment in an American university females had their cere colour artificially manipulated to test the males' reactions. Males were more likely to chose females with a dark cere.

Breeding space

Parrots are incredibly adaptable birds. When we think how different are the conditions of those in captivity to free-living birds, it is surprising that they breed at all. Some do not – until a certain aspect of their environment is corrected. One breeder of Major Mitchell's (Leadbeater's) Cockatoos told me that his pairs did not breed until he moved them to larger aviaries. Then the female of one pair bred for nearly ten years, producing clutches of five, four and three eggs.

In the breeding centre of Palmitos Park, Gran Canaria, there were five pairs of this cockatoo in individual aviaries that were far too small – – 2.4m (8ft) x 1.2m (4ft) x 2.1m (7ft) high. They did not attempt to breed and clearly were not happy in the confined space. Then they were all moved together to a very large aviary on exhibit in the park. It was a pleasure to see them moving as a flock, foraging across the turf floor of the aviary. There were no nest-boxes but one pair excavated an ornamental log and hatched three chicks there. In the wild this cockatoo nests as far away from another pair as possible so it was interesting that breeding – by one pair only – occurred.

Courtship feeding

The owner of a Grey Parrot once asked me what his bird was doing when it dropped its wings and started to vomit when it was sitting on his hand. He had failed to realise that his bird was trying to feed him – that it considered him as its mate! Courtship feeding involves the regurgitation of food; this is not vomiting. In some species, such as Greys, the rituals which accompany it are very pronounced, and include lowering the wings and exaggerated head swirling movements. In other species there is little ritual.

Male Stella's Lorikeet feeding the female prior to egg-laying

Courtship feeding varies in its length and intensity according to the species. In some, such as lovebirds, it occurs throughout the year, but especially during the breeding period, and helps to maintain the pair bond. In other species (eg, Amazons) its duration is short, occurring just prior to and during the breeding season.

The strangest courtship feeding I have ever observed is carried out by hanging parrots *(Loriculus).* In fact, it is probably unique among parrots. The male regurgitates a globule of food, which seems to hang at the tip of his beak – or he takes it back, apparently sucking it in again. If the female accepts this offering it might be followed by mating, which can last quite a long time.

Egg-laying

Clutch size in parrots averages about four eggs although the Palm Cockatoo and occasionally black cockatoos lay but a single egg. In Carnaby's Cockatoo, in which the second egg is like an insurance and rarely results in a fledged youngster, the second egg can be laid as long as eight days after the first. The largest clutches are produced by Rosellas and *Pyrrhura* Conures – up to nine eggs. Thirteen eggs were found in the wild in the nest of one *Pyrrhura* conure – but two females might have been responsible.

In Bonaire, an island in the Netherlands Antilles, researchers found that Yellow-shouldered Amazons *(Amazona barbadensis)* that nested near plantations of fruit trees laid larger clutches than other females. This could mean that in a captive situation, certain parrot species lay larger clutches on a consistently plentiful and nutritious diet.

A study of Green-rumped Parrotlets in Venezuela showed that the clutch size depended on how well the female was fed by the male during incubation. In other words, if food was apparently abundant,

Austral Conures with nine eggs and one newly-hatched chick. In sociable species such as conures, it is possible that two females laid in the same nest.
Photograph © Soledad Diaz

more eggs would be laid. This might explain why clutch size varies in some species in captivity. In most lory species the clutch invariably consists of two incubated eggs; if, for example, the first was laid from a perch, a third would be laid.

Birds are classed in two categories: determinate layers, who produce a set number of eggs in a clutch, regardless of the loss or removal of some of them, and indeterminate layers. Parrots belong in the latter category. This means that if you remove a female's eggs as laid, she will continue to lay more. One House Sparrow laid fifty eggs after her eggs were removed as laid.

Why does this happen? In indeterminate layers egg production is regulated through the brood patch on the female's underparts. If eggs are removed as laid, there is no tactile stimulation of the brood patch and no message to the brain to limit egg-laying. When the eggs remain in place, they are detected by touch sensors in the brood patch and, via a complex hormonal process, allow only the normal number of ova to develop in the ovary (Birkhead, 2012).

The brood patch can sometimes be seen when an incubating female leaves her eggs: it just looks like a bare patch of pink skin. It is a very sensitive area because birds regulate the temperature at which the eggs are incubated by increasing or decreasing the flow of blood to it.

Many pet birds, especially Cockatiels, lay eggs despite the absence of a mate, probably stimulated by a rich diet, long hours of light and/or misguided stroking by the owner. The eggs should never be removed as laid because it could result in the female laying an abnormally large number of eggs and depleting her calcium reserves. This can lead to egg-binding and death, or other health issues. For this reason, the female's eggs should be left in the cage, even if she is incubating them on the cage floor, until she loses interest in them.

The artificial conditions of captivity, including an over-stimulating diet, too high in fat, can cause excessive egg-laying which becomes life-threatening. Altering the diet, reducing the hours of artificial light, obviously removing the nest-box if one is present, and putting a grid in the cage bottom or even water, will deter most females from laying. But not all of them. The owner who is prepared for the vet's bill might opt for a slow-release hormone implant, injected into the female under anaesthetic. This will stop egg-laying for several months.

The reverse problem, that of a confirmed female spending much time in the nest-box without producing eggs, is frustrating for breeders. Surgical sexing is rarely carried out these days but surgical investigation can indicate if there are few egg follicles on the female's ovaries, as can happen in an old bird. In a female which offers a good prospect for breeding, eggs can be seen through an endoscope as many white spots on the ovary.

Infertility of some eggs in the clutch is common. In the wheatbelt of Western Australian, about 20% of eggs were infertile for all cockatoo species found there. The same was true for Glossy Cockatoos on Kangaroo Island and Palm Cockatoos in the Cape York Peninsula. In captivity it often happens that the entire clutch is infertile, due to one bird not being in breeding condition. Hand-reared males are often useless for breeding as they do not know how to copulate – so all the eggs will be infertile.

Incubation

The incubation period averages about twenty-five days across all parrot species. It varies from eighteen days in the Budgerigar and other small species up to 30 days in the Palm Cockatoo and other large cockatoos. The exception is the Lesser Vasa Parrot which has the extraordinarily short period of fourteen days. The ambient temperature can shorten or increase the length of time before hatching.

Incubation normally commences with the laying of the first or the second egg, thus there can be a significant size difference between the first and last-hatched young. The advantage of this system is that if there is an inadequate food supply, the last-hatched will die. increasing the chance of survival of the older ones. Each surviving chick receives more food. If, in captivity, the smaller chicks in the nest die, the breeder should question whether he or she is not only providing enough food – but enough *soft* food.

In almost all parrot species only the female incubates. The exception relates to white cockatoos, including Galahs, and Cockatiels, in which males take their full share of incubation. In at least some *Charmosyna* lorikeets incubation is also shared by both sexes. In the *Calyptorhynchus* (black) cockatoos, only the female incubates. Some males of other parrot species spend time in the nest with the female but are extremely unlikely to take part in incubation. However, from captive birds we know (CCTV cameras in nest-boxes) that in rare cases the male will incubate for a short time when the female leaves the nest.

Obviously in the wild, in most species only the female incubates so the female does not have time to forage for food. She relies on the male (or males in the case of Eclectus and Vasa Parrots) returning to the nest to feed her The male calls out to the female when he flies in with a cropful of food. He either enters the nest or the female comes out and sits on a nearby branch (usually the same one) to be fed. In the study of a pair of Racket-tailed Parrots, twenty occasions were observed; the male fed the female inside the nest on eleven occasions and outside the nest nine times (Walker and Seroji, 2000).

In captivity some male parrots seem to have lost the inclination to feed the female during incubation – something that the breeder needs to watch out for. A female might have such a strong instinct to stay in the nest that she could starve to death if the male is not feeding her. Putting a piece of millet spray in the nest-box every day might solve this problem.

Hatching

Once the chick pips into the air space it can hatch in a few hours or take up to two days. It uses the egg tooth – a small projection near the tip of the upper mandible – to pierce the shell. Here I would like to lay to rest an oft-repeated inaccuracy. The egg tooth does not "fall off". It is covered by layers of keratin (the substance from which the beak is made) as the chick grows. The remnants of the egg tooth can be visible for up to two weeks.

In many species, chicks call in the egg up to three days before hatching. In a study of captive Budgerigars cheeping calls were heard one to two days before the chicks hatched. Video recordings were made of females actively assisting hatching up to a few hours before the chick hatched. This happened in all ten cases in which the event was videoed. The hatching chick either rolled out or was lifted out of the shell by a parent. Sometimes an attempt was made to feed the hatchling while it was still partially in the egg or immediately after it was free from the shell. The time between a parent starting to nibble at the crack until the chick was

Amazon parrot chick hatching in an incubator

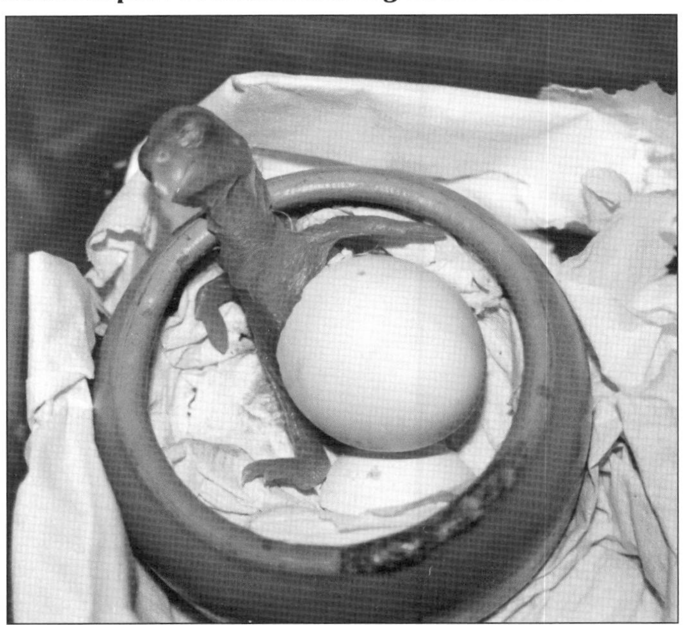

The egg tooth (the small white projection) is visible in this Pesquet's Parrot, photographed within the first hour of its life.

outside the shell ranged from one to twenty-two minutes, and averaged about seven minutes.

In at least three cases, the female gave an audible call when alone and nibbling at the egg. Within moments the male entered the nest-box and both birds then assisted the hatching chick. Clearly, the male had understood the female's call. There was a striking contrast between the behaviour of the male and the female at this time. Females were much calmer in assisting chicks to hatch when they were in the nest alone. Judging by the speed and jerkiness of movement, when the male was present both parents moved around excitedly, seeming to get in each other's way. Sometimes their efforts were simultaneous, probably accidentally, and they pulled the shell apart from opposite directions (Berlin and Clark, 1998).

In captivity, assisted hatching by a parent is not always successful, and can lead to the death of the chick. If an overdue egg must be opened to save a chick's life, the best strategy, in my experience, is to carefully remove about half of the shell, keep the membrane hydrated, then allow the chick to hatch in its own time. This prevents haemorrhaging.

There is still a great deal to learn about the breeding biology of parrots – and doubtless a few surprises to be revealed.

31. REARING THE YOUNG

There is so much that can go wrong while young are being reared. For wild birds, the regular presence of parents at the site can attract the attention of predators, ending the lives of adults and/or young. In recent years loss of habitat means that many parrot species have to forage further away from the nest, leaving it unguarded at times.

For captive parrots, the rearing period can be very stressful, especially if male and female are not compatible or the rearing foods are unsuitable or inadequate. A female might starve herself to ensure her chicks are well-fed and one who is not totally fit might be found dead in the nest at this time.

Parents' tender behaviour

Most parrots show great devotion to their young. In recent years we have started to understand more about what happens inside a parrot's nest. This is because in a number of parrot conservation projects scientists employ tiny nest-box cameras. Many breeders also use this equipment, with monitors in their house, to find out exactly what is going on. Increasingly zoos are using this method to allow visitors a fascinating glimpse into nests.

Everyone who has observed parental behaviour is impressed by the tenderness and gentleness with which parrot chicks are fed, preened and brooded. Feeding is especially interesting. Take a macaw, for example, with its massive beak, and consider what great care it must use in feeding a tiny newly-hatched chick whose beak is only one centimetre long.

In the USA Rita Groszmann mounted an infra-red camera on the nest-box of her Green-winged Macaws so that she could observe what was happening, night and day. The single chick was fed six hours after it hatched. "Carefully and tenderly, she reached for the tiny, soft, little beak and gently using her lower mandible like a bowl, scooped regurgitated food out with her tongue into the baby's mouth."

The male's manner of feeding the chick was less smooth and less confident; but after feeding it he would tenderly nuzzle, lick and gently cuddle it against his breast feathers. After the chick's eyes opened, it would reach for both parents, begging for food.

Rita Groszmann wrote: "As a one-week or younger baby, there was a great deal of tongue action controlling the flow of food into the young chick's mouth. At two and one half weeks of age or so, the baby's eyes opened and the feeding response became very strong, almost violent, accompanied by wild flapping of naked wings." (Groszmann, 2002).

These observations should help some breeders to understand that to take chicks for hand-rearing causes the parents, especially the female, a sense of loss and even sorrow. It is not the same as removing young when they are fledged and independent. Eventual separation is normal – although it usually occurs much earlier in captivity.

There should surely be less emphasis on hand-rearing and more emphasis on producing parent-reared young for future breeding stock. The continued removal of eggs for artificial incubation or the removal of chicks at an early age must have detrimental psychological effects on breeding pairs. It interrupts the breeding cycle which, if allowed to be completed, keeps the pair occupied for several months in the most natural way possible. This surely is the best form of environmental enrichment!

Another aspect of removing chicks for hand-rearing, that of the life-long psychological damage of young parrots raised away from their own species, is emphasised elsewhere in these pages.

Do parrots feed their chicks during the night?
The answer is yes – at least some do. This can be difficult to detect in many parrot species as the sound of chicks being fed is not loud – unlike that of *Cacatua* cockatoos. In a study of Slender-billed Cockatoos *(Cacatua tenuirostris)* in western Victoria and South Australia, observations were carried out at two nests. Sounds of chicks being fed were heard as late as 11pm and it was suspected that feeding continued throughout the night. In the 1970s I kept a pair of the closely related Goffin's Cockatoos in an aviary quite close to my bedroom window. I can confirm that they, too, fed their chicks during the night! Some people who hand-rear parrots believe that it is not necessary to give nocturnal feeds because the parents don't do it. It is usually true unless chicks weigh less than about 10g or have a health problem.

Chick mortality
Causes of deaths of parrot chicks in wild nests are varied. They include predation by birds of prey, hawks, rats, stoats and snakes. An interesting study was carried out on the Vulnerable (actually down-listed from Endangered, but still declining) Hyacinthine Macaw in 1995 and 1999-2003. It showed that at 346 nests, 138 eggs were predated. Seventy-one were taken by toucans, 26 by jays, 12 by opossums, four by coatis and 25 by unknown predators (Pizo et al, 2008). This situation is somewhat ironic because 95% of Hyacinthine nests in the Pantanal are in *Sterculia* trees – the seeds of which are dispersed by toucans!

In Some Australian cockatoos the rate of predation is even worse. On Kangaroo Island brush-tailed possums took 40% of all eggs laid in accessible nests of Glossy Cockatoos *(Calyptorhynchus lathami).* On the Cape York peninsula, 24% of eggs in Palm Cockatoo nests studied by researchers were predated by goannas, large rodents and Black Butcherbirds.

For many parrots humans are by far the most serious predators, taking all the young and never allowing any to fledge. This has brought some species (such as Amazon parrots in Mexico) close to extinction.

Other causes of chicks mortality are flooding of nests and infestation by parasites and fly larvae (See **33. Bacterial and Parasitic Infections in Chicks**). Shortage of food will, of course, cause chicks to die. Creatures taking over nests, such as bees and even parrots of the same species desperately looking for a nest, and killing or throwing out the young, can also be to blame. In Uganda, a nest of Grey Parrot chicks was killed by a pair of Black and White Casqued Hornbills. They disposed of the bodies and took over the nest site.

Cockatiel with newly hatched chicks. Parrots are caring and protective parents.

Photograph: M. Scrivener and A.Chaloner

In aviculture

In captivity chick deaths are often associated with poor management – incorrect diet, disease (especially viruses and bacterial infections), infested, dirty or damp nest-boxes, mice and other vermin. Too frequent nest inspection could cause parrots to kill or desert their young.

Low temperatures in pairs breeding during the winter cause chick deaths because different species automatically cease to brood chicks during the day (regardless of temperature) at a certain age – related to their habits in the wild. At least one breeder solved the problem with the use of a heated reptile rock in the nest-box. Great care would need to be taken that no electric cable was exposed. It would need to be enclosed in a strong tube.

Disturbance beyond the control of the breeder, such as storms and other severe weather, and loud or prolonged unusual activities in the neighbourhood, can cause parents to desert the nest.

When pairs are rearing young, especially if there are more than two chicks, the rearing food needs to be replenished several times daily, and other soft or fresh foods increased in quantity. If the parents are not offered an adequate amount of suitable (not hard) foods, the youngest chicks will probably die.

Aggressive birds in an adjoining aviary can be a major problem. As already described, in the wild many parrot species occur in flocks after the breeding season or when they meet at a good food source. These are the sociable species. Others are seen only in pairs or family group. Examples are Hawk-headed Parrots and caiques. They do not readily tolerate their own species in the vicinity. This is why breeders need to learn about the breeding biology of the species they keep.

Those who house two pairs in adjoining aviaries often encounter such problems as chicks being killed in the nest, due to displaced aggression. The male cannot kill the neighbouring male which he perceives as being on his territory so in frustration

Goldie's Lorikeet chicks – a male and a female. Experienced breeders can often sex chicks in the nest by physical features such as head shape and size.

he enters the nest and kills a chick or chicks – or even his female. Or perhaps one male is intimidated by the close presence of another, so only one pair will breed.

It is not unknown for some male parrots to kill young of the same sex as themselves soon after they fledge. Whatever the reason (potential competitors?), in known cases the male should be caged within the aviary just before the young fledge and not released until they are independent and have been removed. Females can usually cope alone unless there are four or more young.

Sex ratio of young

Parrot breeders often bemoan the fact that they produce more males than females. The demand for females is always greater. But what happens in the wild? Are more males produced? Once again we have to thanks Steve Beissinger for the only study of parrots that has tracked hundreds of individuals of a sexually dimorphic species. Over 14 years he and his research team looked at the sex ratio of 2,728 nestling Green-rumped Parrotlets in 564 broods. Fifty-one per cent of young hatched were males.

Several studies have shown that mortality is higher in females than in males. Tagged Major Mitchell's Cockatoos revealed that the average annual survival rate of adults was 93% in males and 81% in females. This indicates a high mortality rate in females. As both sexes incubate and care for the young, predation in the nest cannot be the reason.

In the New Zealand Kaka *(Nestor meridionalis)*, the sex ratio on the mainland was three males to one female, almost certainly because stoats killed many incubating females. This is a very worrying situation in a declining (and incredibly charismatic) species. At one nest studied two males seemed to be in attendance while at another three males were close by. On Kapiti Island, where predators have been eradicated, the sex ratio was 1:1 (Greene and Fraser, 1998). This underlines the high mortality suffered by defenceless female in the nest.

Communal breeding/nest sharing

While some parrots nest in loose colonies, I use the term communal breeding to mean the situation where several birds share the same nest site. This is common in *Pyrrhura* and some other conures (parakeets) in the neotropics. The groups probably consist of a pair and their young from the previous year, or perhaps two years, and possibly one or more unrelated individuals.

When I decided to try to recreate this situation in captivity, I left the two 2013 young (a male and a female) with their parents. The female laid seven eggs in 2014. Six chicks hatched but two died soon after hatching. During the rearing period the 2013 young continued to enter the nest-box at night – just as happens in the wild. At no time did they enter the nest to feed the young. When the four young fledged it was touching to observe the tender care – preening them and fussing over them – provided by the older offspring. One was also seen to feed a young sibling who approached for food, dropping its wings and rapidly fluttering them, while making plaintive food soliciting sounds. It will be interesting to find out how long this group can live together in harmony.

Few studies have been made, as yet, to determine if the chicks in nests where several adults are present all have the same parents. DNA finger-printing was carried out in captivity on the Endangered Golden Conure *(Guaruba guarouba)*, another communal breeder. In one nest with five chicks it was discovered that four chicks were fathered by one male and one chick by another (Albertani, Miyaki and Wanjtal, 1997).

Pre-fledging and fledging

The period that young parrots remain in the nest varies from just under five weeks in some small Australian parakeets (such as Budgerigars) and *Forpus* parrotlets to about sixteen weeks in the Hyacinthine Macaw. For most medium-sized parrots it is about eight weeks. They are then fully feathered but the tail feathers are shorter in most species and the colours duller.

Maximum nestling weight is usually achieved about half way through the rearing period, or up to two-thirds of the way. The weight then reaches a plateau. Chicks are fed less often as fledging approaches and lose some weight before they leave the nest – or they would be unable to fly. At this stage they will vigorously flap their wings within the nest, no doubt to strengthen the wing muscles. Most can fly strongly on fledging. Some will stay quietly in the vicinity of the nest, making their food begging calls and waiting for the parents to fly in and feed them. At this stage they are extremely vulnerable to predation.

Occasionally a young parrot will leave the nest prematurely – perhaps accidentally in its eagerness to be fed at the nest entrance. I once spent a few memorable days on the Murray River in a houseboat with friends. One morning, as we passed a riverside eucalypt, a young Little Corella fell from its nest. We drew to a halt, climbed on to the bank and saw it sitting in low, spiky vegetation. My companion said that a fox or a hawk would take it by nightfall. It looked as though it was nearly ready to fledge but I doubted that it would be able to fly. Its fate haunted me that evening as I wondered whether its parents had managed to coax it upwards. It looked so vulnerable and helpless...

32. WEANING AND INDEPENDENCE

Information on how long parrots stay with their parents and are fed by them is scarce. This is especially the case in heavily wooded habitats where observation of parrots in the canopy and the identification of individual birds is difficult – or just impossible.

The Galah was written by Ian Rowley, a scientist who studied *Eolophus roseicapillus* for years before his 190-page book was published in 1990. This was one of the earliest in-depth studies of a parrot species in its natural habitat. His observations indicated that young birds remain entirely dependent on parental feeding for six or seven weeks after leaving the nest. This parrot is one of the few in which young birds form crèches. He stated: "Within a month of leaving the nest, the young Galahs start gathering grain for themselves" and that they cease to be fed by their parents when they are about 100 days old, which is about two months after they leave the nest.

Many young parrots are taken by predators, probably mainly birds of prey, in spite of the fact that they fly very strongly soon after leaving the nest. Rowley tagged 154 Galahs as nestlings. He found that 19% died between the time they left the nest and when the parents left them at about 100 days old. A further 22% of those birds that reached independence died during the ensuing two months. Only 49% of the original 154 survived to the age of six months.

Another study of cockatoos in the wheat belt of Western Australia was made by Denis Saunders who studied the nesting habits of the Red-tailed Black Cockatoo. The single youngster spends three months in the nest and is fed by the parents for a further four months. On occasions a female will lay another clutch after about two months of fledging a youngster (Saunders, 1977). If this occurred, presumably only the male would feed it; he would not wish to prolong this duty as he would also be feeding the female in the nest.

Saunders also studied Carnaby's Black Cockatoo from the same region. Similarly, the young spent about three months in the nest and were independent at about seven months old. The young bird would remain with the parents until the next breeding season, or even longer.

Studies such as those by Rowley and Saunders are extremely rare but possible to carry out in the open habitat of the wheat belt. In the Greater Sulphur-crested Cockatoo *(Cacatua galerita),* for example, it is very difficult to gain this information as families leave the nesting area soon after the young fledge to join post-breeding flocks.

It is the same with some Australian parakeets, such as the Barraband's (called Superb Parrot in Australia). Parents and young also leave the breeding area, first to nearby woodlands, and then much further away. This makes continued observation impossible. The large *Platycercus* parakeets such as the Crimson Rosella are reported to be fed by their parents for three to four weeks after vacating the nest (Forshaw, 2002).

In captivity, breeders usually leave parakeets with their parents for three weeks after fledging, or perhaps only two weeks in the small *Neophema* species, by which time they are fully independent. However, if a family group is compatible, it is beneficial to the young of all species to remain with their parents for much longer, as they learn so much from them.

Cockatoos and other large parrots are reliant on their parents for several weeks because they need to learn how to manipulate and open seeds and nuts. In contrast, the lorikeets' diet of nectar and pollen is easy to harvest and to digest. Reportedly, small species in Australia, such as the Purple-crowned *(Glossopsitta porphyrocephala),* are independent one week after leaving the nest. In captivity hand-reared lories and lorikeets learn to feed themselves and are sometimes fully independent before they would even have fledged, had they been in the wild. When parent-reared, however, they may solicit food for some days after fledging, but very quickly attain independence.

In Brazil, the largest macaws, such as the Hyacinthine, are believed to spend up to one year with their parents. This may be partly due to the fact that it takes them a long time to learn to open the nuts which form the large part of their diet. Strength of the bill may also be involved as on fledging it is not as hard and powerful as that of an adult. In Bolivia, the smaller Red-fronted Macaws were observed being fed by their parents when they were at least eight months old. In South Africa, Cape Parrots are feeding independently at the age of one hundred days (Perrin, 2012).

Hand-reared parrots

Sadly, the majority of breeders wean parrots much too early. This has a profoundly negative effect on their future and well-being. It applies especially to some white cockatoo *(Cacatua)* species and to the large macaws. The only exception to this weaning age among the cockatoos is that of the Australian species such as the Bare-eyed (Little Corella), also the Galah, which weans much earlier. Their temperament is different, much more independent than the large species such as the Moluccan and the Umbrella that are usually sought as pets because they are so "cuddly".

Many breeders sell these species before they are fully weaned, at 15 or 16 weeks or even before. In fact, nearly all hand-reared parrots are sold too early because weaning at the bird's pace, not that of the breeder, takes a long time, is time-consuming, needs patience and cuts profits. The buyer is told that the young parrot is weaned, or that it might need one feed a day for a couple of weeks.

Many parrots are forced-weaned, that is, made to eat on their own before they are physically and emotionally ready. The process does not only involve eating food on its own, it also involves breaking the bond that has developed between the person(s) who fed it and the parrot. The double impact of this and being forced to eat on its own, usually results in a young parrot that is extremely insecure and constantly begging for food and crying for attention. The problem is compounded by offering them unsuitable hard foods – an assault on their digestive systems. They need soft, easily digested foods for many months. Such unfortunate young parrots are liable to have behavioural problems all their lives – assuming they survive the experience. Sadly, many purchasers are too inexperienced to know that the young bird is starving.

Advice that I have heard a hundred times is that the purchaser of a hand-reared parrot should not continue to hand-feed it because this discourages it from eating on its own. This is the most unsympathetic and cruel advice imaginable. The fact is that when a young parrot is hungry and anxious, it will cry, beg and wing-flick incessantly, making no attempt to feed itself. When its crop is full, or nearly so, it will start to nibble at food because the anxiety is gone. The caring and sympathetic breeder will always hand-feed a parrot as long as it asks – no matter how long this takes.

With the exception of two sibling conures whose development I needed to monitor and record, it is some years since I hand-fed a parrot. Unless hand-reared birds can be socialised with their own species as soon as they are independent, hand-rearing does permanent harm to their natural instincts and behaviours. I no longer advocate it.

33. BACTERIAL AND PARASITIC INFECTIONS IN CHICKS

Research on diseases affecting free-ranging parrots is scarce – but increasing. Bacteria are found naturally in most living organisms. The normal intestinal flora of parrots consists predominantly of a group known as Gram positive, such as lactobacillus and staphylococcus. These bacteria take days or weeks to colonise the tracts of chicks, passed to them by the parents or other organisms with which they have contact. The natural intestinal bacteria aid in the digestion of food and protect against the growth of harmful bacteria.

Infections caused by Gram-negative bacteria are common in captive parrots; these micro-organisms are considered either pathogenic or opportunistic. One such bacterium, *Escherichia coli,* lives in the digestive tract of animals and humans. It is frequently involved in respiratory, digestive and septicaemic disorders in captive parrots. Baytril (enrofloxacin) is widely used by veterinarians for treating bacterial infections but it must be remembered that the use of antibiotics weakens the immune system.

In Brazil cloacal samples were taken from wild nestlings of Blue-fronted Amazons and Hyacinthine and Lear's Macaws to determine if they could be carriers of recognised *E.coli* pathotypes. A total of 44 samples were obtained: 21 from Blue-fronted Amazons, 13 from Lear's Macaws and ten from Hyacinthine Macaws. *E.coli* was obtained from all 44 samples. All the nestlings in the study fledged, indicating that although the potential for disease was present, birds living in their natural environment are more likely to remain disease free. It was thought that several factors most likely to be found in captivity needed to be involved in triggering disease development (Saidenburg *et al,* 2012).

Also in Brazil, health monitoring of chicks of Blue-fronted Amazons included testing for chlamydophila (psittacosis). Cloacal swabs showed that 6.3% of young tested were positive; in tracheal swabs the figure rose to 35.6%. All looked healthy. The disease was not a problem in free-living birds – but when such parrots were trapped and experienced the cruelty and stress of trapping and captivity, countless thousands died.

Some veterinarians prescribe antibiotics for chicks in aviaries found to have *E.coli* and other Gram-negative bacteria when their faeces are examined. However, such medication often does more harm than good, in killing the normal bacteria in the gut. Medicating chicks should surely be a last resort for chicks that are obviously sick only.

Parasites

The risk of parasitic infections is greater in nests that are reused for several seasons, such as the domed stick nests of Quaker (Monk) Parakeets. In Argentina, fifty-two chicks of this species, aged between three and forty days, were examined for parasites. Twenty-six were infested. A single species of parasite (mites or lice) was found on nineteen nestlings and two species were found on six nestlings. Three species were found on only one bird.

The nests of Quaker Parakeets, such as this one in Brazil, harbour many parasites.

When I worked and lived in Loro Parque, Tenerife, Quaker Parakeets nested in the palm trees above the terrace of my apartment. It could happen that a stick nest became so large that it fell down. On one occasion I rescued the eight chicks, which were hand-reared, then given away as pets. First, however, they had to be treated with insecticide powder to kill the red mite which were crawling all over them! And me – as I picked up the chicks!

In the Pantanal region of Brazil, veterinarian Dr Mariarigela de Costa Allgayer monitored the health of Hyacinthine Macaw chicks. In blood samples from 112 nestlings from 85 wild nests no blood parasites were found but several chicks (3.3%) were infested with ectoparasites (*Philornis* flies). Some had scars on the head and wings, proof of recent infestations that could delay fledging by ten days.

In Argentina, during the long-term study of the Burrowing Parrot (Patagonian Conure) colony at El Cóndor, Juan Masello and Petra Quillfeldt made an interesting discovery. While examining and weighing nestlings, they found fleas in the nostrils and under the tongue – apparently a unique phenomenon among fleas. Those collected from more than two hundred chicks turned out to be a new species. In Guatemala chicks of Yellow-naped Amazons studied by researchers had suffered fly larvae wounds on the forehead. Others were infested with lice.

In aviaries

Captive parrots can also carry parasites, the best known of which are intestinal worms. They mainly affect aviary birds, especially parakeets, which have access to earth floors. Their presence can be easily detected by examining faeces for eggs under a microscope. Once it was appropriate to worm some newly imported parrots which carried tapeworms. These days, with no legal importation of wild-caught parrots, tapeworms are rarely seen. However, incidences of parasitic infections, such as avian malaria *Haemoproteus* blood parasites, may have increased.

Signs of parasitic infections of all kinds include brittle feathers, diarrhoea, blood in the faeces, dehydration, weight loss, thirst and excessive food intake. Another symptom – seldom recognised – is when a bird loses the use of its legs. It is advisable to consult an avian veterinarian if these symptoms are observed.

High humidity and warmth encourage parasitic infections; vermin, including cockroaches, carry parasites. Seed can also be a source. Note that some disinfectants are effective against parasites. Neem is a safe natural organic insecticide that can be used to treat chicks in the nest.

PART VI.
FOODS

Yellow-crowned Amazon Parrots foraging – wild and captive.

34. FORAGING AND FOOD

In their natural environment most parrots spend between one quarter and one third of the daylight hours foraging and feeding. The time spent depends partly on the energy values of the foods taken and how far they need to fly to find them. In Australia, rainfall will significantly affect foraging time. One pair of Galahs spent eleven hours in this activity on one day when plant growth was minimal. One month later, after heavy rain, the same pair spent only four hours per day searching for food.

With the exception of a few specialist feeders the food they consume might vary throughout the year – not necessarily on a daily basis but as they exploit what is available. In forested habitats it might be that a large fig tree provides so many fruits that they feed there for several days. Then the bonanza is over and they must forage more widely, perhaps not finding another fruiting fig but consuming a variety of small fruits and berries until another tree is at the fruiting stage.

Parrots that rely on fruiting trees seldom occur in big flocks – although large numbers may gather at a good food source. In contrast, others consume foods which can be abundant but are usually widespread, such as seeding grasses. Budgerigars and parrotlets are examples. They often travel together in large flocks, the main advantage being that the individual's chance of locating food is increased.

Depending on the species, the season and the weather, the time spent locating and consuming food varies from day to day and might be highly unpredictable. This is a parrot's major preoccupation (except for females incubating eggs or brooding young chicks). It keeps them very busy! However, one study of Quaker (Monk) Parakeets showed that they spent 22% of daylight hours foraging and 27% nest building and repairing.

Parrot specialists, such as Lear's Macaws, might need to fly 30km (18 miles) from the roost site to find licuri palms with ripe fruits. Having found them, they can get all the nutrition they need in a two-hour feeding session in the morning and another in the evening. But reaching the locality is time-consuming.

A study by Alexander Brandt in the early 1980s indicated how foraging was related to the time of year. In July, for example, when few palm fruits were available, 26% of daylight hours was spent foraging on corn crops (it took 20 minutes to eat a full ear of corn) and 6% in palm trees. In September, when no corn was available, 36% of their time was spent searching for palm fruits. In both cases this occupied one third of their active hours.

In South Africa, the Critically Endangered* Cape Parrot, another species with specialised needs, is known to travel up to 100km (63 miles) per day to find food. The fruits of yellowwood trees (*Podocarpus* species) are important but are now hard to find due to reduced areas of yellowwood forests and little natural regeneration.

---

* Critically Endangered, as indicated in the South African Red Data Book of Birds. Fewer than 1,500 individuals survive in the wild.

In contrast, pollen and nectar-feeding parrots such as lorikeets, are nomadic out of the breeding season and stay in an area which has abundant blossom of the right species. Unlike other parrots, they must feed throughout the day, because their food is digested rapidly.

Fruit-eaters

Primarily fruit-eating parrots, such as the Lesser Vasa Parrots *(Coracopsis nigra)* from Madagascar, feed mainly (about 70% in this case) on unripe fruits. This allows them to take the fruits before other birds and animals can eat them. Not all captive parrots like soft fruits so it is a good idea to sometimes offer them quite hard fruits that are not fully ripe, to test their reactions. For example, raspberries and strawberries which are, to our taste, not quite ripe, are readily eaten, unlike those that we prefer.

In Europe captive parrots are usually offered commercially grown fruits such as apples, oranges and grapes whose nutritional contents are not high, but they do contain some valuable properties in very small amounts. The most healthy fruits and vegetables are those high in beta carotene (converted to Vitamin A) such as mango and papaya – but they are expensive and therefore not usually fed on a regular basis by anyone with more than three or four birds. Two other nutritious fruits are pomegranates – a universal favourite – and apricots, if the birds will eat them. Mango is very widely eaten by parrots throughout the tropics – and many consume it before it is ripe. Note, however, that items containing beta-carotene lose up to half this content when stored in a refrigerator. If possible, store outside the 'fridge.

All the fruits and vegetables that we buy from supermarkets or markets have a very low protein content but a high sugar content. Sugar is a good source of energy but for companion parrots which have little chance to burn this off, this should be borne in mind. The dried fruits, such as pineapple and papaya included in some parrot mixtures, are a pointless and inadvisable ingredient with a very high sugar content.

Fruit and vegetables are valuable for their fibre content, even although they are not the easiest food for parrots to digest; friendly bacteria in the gut are needed for this process. Certain species, such as Eclectus and Amazon parrots, have evolved longer hind guts to cope with fibre digestion, and for these species a higher level of fruit in the diet is appropriate. Feeding fruit to our parrots assists bowel function; it might also help to redress some of the nutrients missing from seed diets. An important point is that it reduces the level of potentially harmful items eaten, such as fatty seeds.

Flowers

In South America, flowers play a large part in the diet of parakeets (Low, 2013). This is true for many parrots – a fact that is unrecognised by most parrot keepers. In Madagascar Lesser Vasa Parrots have been observed consuming about sixteen flowers per minute. In 10% of the plant species consumed, only the flowers were eaten (Bollen and Van Elsacker, 2004).

Flowers are used as food by many parrot species, consuming most of their parts. One study of Grey Parrots in Ghana showed that although they feed mainly on fruits and seeds, flowers constitute about 20% of their food (Tamungang and Ajayi, 2003). In captivity, they relish the blossoms of the easily cultivated nasturtiums – among others. Those of hibiscus, roses, pansies, marigolds, and blossom from fruit trees, will also be enjoyed. They are important to provide natural foraging, colour and amusement, even if the parrots eat little.

Nuts

Nuts have a very high energy content. Depending on the season, many wild parrots must fly long distances, thus they use up a lot of energy. Our birds use far less energy so their daily allowance of nuts should be rationed, except when used as a stimulus to breeding. However, for macaws and other large parrots, nuts are more than the sum of their nutritional value. Opening nuts in the shell is the most natural action they can perform. Some parrots also crunch up the shells, providing another way of keeping them busy. It can be a lot of focussed and enjoyable work for smaller parrots to open nuts.

It is difficult to find nutritional analyses for nuts on which parrots feed unless they are also used for human consumption as are the kernels of fruits of the licuri palm. They have a fat content of about 49% and contain 11% protein. In addition, they are high in Vitamin A or, more correctly, beta-carotene, and they also have a high calorie content.

Foraging branches

Enrichment in the form of fresh-cut branches or edible items presented in an interesting way is usually the most effective form of enrichment. In Europe willow and apple branches are favoured, or hawthorn with blossom, whereas in Australia eucalyptus and *Casuarina* are widely offered.

My conures relish sprays of dock *(Rumex crispus),* preferably fresh – but dried in the winter. They take the seeds by hanging head downwards while perched on them. I am giving them a natural foraging experience – and giving myself the immense pleasure of watching them enjoying it. They love dried figs, briefly soaked to make them more supple, and they take these from the fruit hangers made for parrots. Yet if I place a fig in their food dishes, it is ignored. They enjoy food that is hung up, or even thrown on the aviary roof.

They would spend all day eating leaves and shredding fresh-cut branches or working on seeding grasses or branches with berries. Natural items will be nibbled, destroyed or eaten but, unlike many larger parrots, most man-made toys are ignored. They prefer natural items.

In 2005 Gay North wrote in *Good Bird Magazine:* "I've often given my birds clean branches in the past so they could rip and shred the leaves but on this occasion, I took it a step farther. I made Foraging Branches. I used raffia to tie individual peanuts to the branches. I made up little cups of veggies to hang. I tied some carrots and hung them. I cut some zucchini into large pieces and then inserted grape pieces into the flesh. When all was done, I fastened the branches to a hanging stand that three birds (Goffin, Meyers, Red Bellied) had free access to."

Almost immediately they started to forage for the food items and when these were eaten they continued to play with the leaves and branches. This is just a start. Time and creativity is all it takes to enormously enhance a parrot's day.

Blue-winged Parrotlet in Brazil feeding on the dried heads of sunflower. These make excellent foraging items for captive birds, hung in cage or aviary.

A word of caution, however. A parrot that is not used to new or frequently changed items within its cage might be wary and suspicious at first. It is therefore advisable to put them within view of the cage, and gradually move them closer. When the parrot no longer appears nervous of the new item, it can be placed in the cage.

The seven Greys belonging to a friend have the best of both worlds. Indoors they have their own cages. Weather and temperature permitting, they are taken outdoors and placed in a large aviary for several hours daily. The aviary is overgrown with herbs, with tall mint dominating. They enjoy foraging among the plants and the shade they provide. They have even been seen on the aviary floor – eating mint! This type of captive foraging is difficult to surpass!

Foraging enrichment

Compare the hours of foraging mentioned above with a captive parrot's situation. Food is in front of it, usually all day. It does not need to work to get that food – only to pick it up and eat it – a monotonous activity. The time spent is very short. Unfortunately, most people caring for parrots offer the same food, in the same place, day after day. In recent years, however, an increasing number are realising the value of foraging toys. There are many kinds on the market. They make life much more interesting for the parrot as it "works" to remove the food, raising its level of activity and motivation, and increasing the time it takes to find and consume its food.

A good starting point for foraging toys is the stainless steel fruit hanger, made for parrots and available through many outlets that cater for birds. Fruit hangers will swing like small branches in the wind. When a parrot is used to taking fruits and vegetables in this way, lots of new ideas can be tried.

One purpose of Yvonne van Zeeland's thesis, *The feather damaging Grey parrot: An analysis of its behaviour and needs* was to determine how effective foraging enrichment is in increasing feeding time. She described eleven types offered to Grey Parrots, such as placing pellets in puzzle feeders. Video recordings were used to analyse foraging times and activities. Some of these devices are described here to encourage parrot owners to provide similar opportunities.

- Four transparent plastic cups with lids, dangling from a PVC pipe, hung from the roof of the cage.
- Transparent acrylic wheel, diameter 15cm (6in). Parrots needed to spin and turn the wheel to access food via a circular hole (2.5cm diameter).
- Honeycomb transparent acrylic feeder (8cm x 18cm) containing a cardboard box filled with food. The birds had to shred the box to reach the food. (The latter two are commercially available parrot foraging toys.)

In a captive environment, finding and consuming food may take less than one hour and is usually too predictable and easy. Foraging toys or puzzle feeders appear to be the most effective measures to increase activity, stimulate, alleviate stress and boredom and reduce and prevent aggression. Foraging enrichment is a wide term which might include using multiple bowls or mixing food with inedible items – but less effective than the toys.

Yvonne van Zeeland and her researchers studied eleven healthy and ten feather-damaging (FDB) Greys on the assumption that the latter would feed from bowls in preference to feeding from toys. This assumption was proved to be true. FDB parrots spent approximately 21% of the feeding time foraging while in healthy birds this was approximately 50%.

Contra-freeloading means giving animals the choice between free access to food or making them "work" for it. In the Netherlands Winny Weinbeck fills the cardboard centre from a roll of kitchen paper with various small items of food, wrapped up in paper. When placed inside its cage, her Grey immediately explored this, removed the items and ate them. It always "forages" in preference to taking items out of the food bowl.

She has dozens of parrot foraging toys for her parrots. Faced with a difficult puzzle, the Grey would never give up until it had worked out how to get at the food, whereas the Galah and the Amazon lost interest if the task was hard. This underlines the importance of stimulating tasks for captive Grey Parrots in particular.

Agile and acrobatic canopy-feeding parrots greatly enjoy more challenging ways of obtaining their food. However, small parrots such as Budgerigars and other grass parakeets *(Neophemas)* and Cockatiels, feed primarily on the ground. They do not use the foot as a hand when feeding, and will show less interest. After all, they are less used to feeding from moving branches.

Sharing mealtimes

Parrots are, of course, social feeders, that is, in the wild, a number will be sharing the same food source. If your parrot is in the same room when you or the family are eating, don't be surprised if he becomes noisy! He wants some of your food! This is a good opportunity to introduce him to items that he might normally refuse. If you put chickweed or hawthorn berries on your plate especially for this purpose, he is more likely to eat them because he believes they are part of your meal. Or you can give him healthy, low-fat, salt-free foods such as cooked pasta and baked beans in tomato sauce, cooked chicken, cottage cheese and hard cheese, hard-boiled egg and, of course, vegetables – preferably steamed because micro-waving destroys valuable nutrients.

35. AGRICULTURAL CROPS – COPY IN CAPTIVITY

Throughout the tropics parrots feed on heads of maize, grains, sunflowers, peanuts and cultivated nuts when the opportunity occurs – often in badly spaced crops where they can land easily. Plantations and orchards are other obvious food sources.

Feral parrot populations, especially those of Quaker (Monk) Parakeets and Ringneck Parakeets, are established in many cities and countries worldwide. Reports of what they feed on are interesting because it shows their preferences, especially in urban areas. This means that we might broaden the diet of some parrots in captivity by offering these foods.

In Orange County, California, Quaker Parakeets show a preference for fruits including oranges, apricots, figs, apples, persimmons, plums and passion fruit. Note that the buds of certain trees (Chinese elm and orchid) are also included in their diet. When fruits are scarce, Quakers have been observed gleaning weed seeds on the ground, including those of puncture vine *(Tribulus terrestris)*. This herb has been used in the traditional medicine of China and India for centuries. These days it is mainly used to improve the performances of sportsmen and human male fertility. This is somewhat ironic because if there is one species which is globally almost ubiquitous and certainly free-breeding, it is the Quaker!

In Argentina, Blue-fronted Amazons are treated as pests in orange groves. They attack the fruits to remove the pips. In the highlands of Costa Rica conures feed on apples. They almost certainly take the pips as well as the pulp. Certain conures and other captive parrots relish apple pips and seek them out before eating the fruit. It is an oft-repeated myth that apple pips are harmful due to their cyanide content! But the quantity is too small to be injurious.

Throughout the tropics many parrot species are considered to be crop pests, the best known examples being white cockatoos in Australia. Some crops are destroyed by cockatoos arriving in their hundreds or thousands. Because the birds are common, killing them has little impact on their numbers. It is a different story in Argentina where the Blue-winged (Illiger's) Macaw *(Primolius maracana)* is almost certainly extinct. Although habitat loss has been cited as the main cause of its disappearance there, this apparently is not the case. During 779 days of field work, researchers came to the conclusion that one of the fundamental causes was lethal control by farmers. This macaw, classified as Near-threatened, is widely distributed in Brazil but is nowhere common.

In Venezuela, in the Henri Pittier National Park, Brown-throated Conures *(Aratinga pertinax)* and Green-rumped Parrotlets caused damage to several crops: corn, sesame, sorghum and sunflower. The *Aratingas* fed on cultivated crops in 58% of observations as follows: maize 57%, sorghum 39% and sesame (4%). Mangos and guavas were only consumed in small amounts.

The parrotlets, which arrived in flocks of fifty or more, fed on sunflower in 34% of observations, sorghum 12% and on guavas in very small amounts. They also fed on flowers and seeds of various other plants including the kassod tree (*Cassia* or *Senna siamea*), *Gliricidia sepium* and the yellow poinciana (*Peltophorum inerme*) (Albornoz and Fernandez-Badillo, 1994). Note that the seeds and bark of *Gliricidia sepium,* often used as a shade tree, are said to be toxic – thus the scientific name, also the common name *mata-raton* (mouse-killer).

Melon crops

In 1992 I received an invitation to speak at a meeting in Sydney, Australia. Afterwards I flew to Brisbane to be met by my friend Peter Odekerken. We drove for nearly 12 hours to reach the outback area where Major Mitchell's Cockatoos live, arriving after nightfall and finding a motel at St George. Next morning we departed at 6.15am. We had little chance of finding the cockatoos after 9am, when they would be resting in the heat of the day – 32°C (90°F) temperature. The area was arid and the earth was red. The vegetation was mallee with narrow strips of callitris-eucalypt woodland along the roadsides – a favourite habitat of the Pink Cockatoo, as it is sometimes known. With its salmon-coloured head and underparts and gorgeously decorated curving crest feathers, it is the most elegant – and arguably the most beautiful – of all the cockatoos.

Sorghum – grown for grain and for fodder.

At 7.30am we struck lucky with a flock of about forty birds. They were feeding, moving around and displaying to each other with raised crests uniquely coloured: white at the tip, then orange-red with a central band of yellow. After half an hour or so, they took off and flew across the road. Lit by the sun, the underside of their wings appeared to be on fire – and what a photograph that made!

First recorded in the area about 1937, so local resident John Beardmore told me, its numbers were slow to build up during the first 20 years. The seeds of pie-melons and paddy-melons were important food sources. In 1979 a flock of 600 of these cockatoos was feeding on Mr Beardmore's property. In the 24-hectare (60-acre) paddock the melons were so numerous – about 91cm (3ft) apart – that it was impossible to plough it. In a few weeks the cockatoos had completely chewed up the melons and then scratched through old cow dung to obtain seeds that had passed through the cows!

Major Mitchell's Cockatoos feeding in a field of paddy melons.

So this is a clue to what we can give cockatoos in captivity: melon seeds. They can be bought from health food stores and are included in some parrot mixtures. Also in Australia, Galahs and Long-billed Corellas (Slender-billed Cockatoos) feed on wheat, oats and barley crops at the germinating stage. These grains are so easy to grow in small pots in soil or compost and, when germinating or with seeds at the green stage, they provide a highly nutritious, inexpensive and extremely enjoyable treat for our parrots.

Oats do very well in the cool, damp, British climate and can be planted any time between spring and autumn. Buy whole oats and use deep plant pots. Few parrots like dry oats (which are valuable for their high phosphorous content), although they are included in many parrot mixtures, so if your parrots ignore them, remove and plant. Few parrots can resist the resulting green heads.

Also in Australia, Rainbow and Scaly-breasted Lorikeets *(Trichoglossus chlorolepidotus)* will feed on sorghum crops, mature and in the soft, immature stage. In Queensland, 51 Rainbow Lorikeets were killed out of a flock that was causing problems to a farmer, due to the extremely dry conditions, causing a shortage of flowering plants. All had sorghum seeds in the stomach – the largest number in one bird being nearly 500. Most of the material was endosperm (the covering of the seed that provides its nutrition). Unlike most lorikeets these two species in captivity will also relish green oats and wheat in the soft, green stage.

It is of interest that small pieces of bark were present in many of the stomachs of the lorikeets mentioned above.

36. FOOD TOXICITY, SODIUM AND GRIT

Do parrots know instinctively if a certain food is toxic? Wild parrots seem to avoid toxic foods but that might be the result of experience rather than instinct.

Many of us dream of escaping to palm-fringed coral islands, where the sky is always blue and the sea is warm. Sun-lover though I am, for me there would have to be a bigger incentive to travel more than 24 hours to a remote island. That incentive in 1991 was a jewel of a little lory, the Tahiti Blue *(Vini peruviana).* Already extinct on Tahiti, only one island in the lory's range was easily accessible in the early 1990s. Described as the most beautiful in the Pacific, Aitutaki fulfils everyone's dream.

This oblong atoll is only 18km² (7 square miles) in area. You can drive round it in ten minutes, admiring the sparkling turquoise and deep blue waters from the coast road. Only one small strip of native forest survives. I booked into the Aitutaki Lodges, famed for their little blue residents!

"You little sweetheart!" I found myself saying out loud when I stood on my balcony and a Tahiti Blue Lory flew fairly low overhead, so that I could observe his little red legs and beak and white bib. I had quite a few sightings that first day. How beautiful they looked as they flew off! They were so vocal I knew at once if they were near, even if their glossy dark blue plumage was hidden among foliage. One of the most unusual of all little parrots, they measure only 15cm (6in) long and weigh about 30g.

Next morning the sky lightened at 6am, with small stormy grey clouds on a backdrop of soft orange. At 6.20 a single lory flew overhead. Seven minutes later there was a pair in the garden. At a hibiscus hedge in a nearby grassy lane, I had fourteen sightings in one hour. It seemed like a paradise for the tiny blue parrots but apart from the hibiscus and bougainvillea around homes, not much was in flower.

I asked Aileen Blake who owned the lodges about any fluctuation in numbers. The decline commenced twenty years previously, she said, when spraying of bananas commenced. Soon after she found all her bees dead. The banana expert from Jamaica, who had initiated the spraying, asked: "Well, what do you want? Bees or bananas?"

The plantations, whose flowers were formerly an important source of pollen and nectar, had become potentially lethal for the lories. However, they had learned not to feed on the banana flowers in the plantation near Aileen's house. She had planted bananas around the garden, for their benefit, and it was there that they fed. So – it seems that wild parrots can learn to avoid toxic substances.

Avocado

In some instances relating to avocado, captive parrots do not know that it is poisonous. I recalled how the Quaker (Monk) Parakeets that bred in the grounds of Loro Parque, Tenerife, never went near the avocado trees. That was good enough reason for me to exclude avocado from the diet of my own birds.

Avocado contains persin which is toxic to some animals. It apparently occurs in two variants, but only one of them is toxic. Different degrees of persin at various stages of ripening would explain the frequent observations from Central and South America where wild parrots consume avocado quite commonly. Apparently they learn from older individuals which fruits they can consume and at which stage of ripeness. As the degree of ripeness changes, their colour alters markedly in the UV spectrum, which is visible to parrots (see 5. **The Eyes**).

The three avocado sub-species *(Persea americana americana, P. a. drymifolia* and *P. a. guatemalensis)* have been crossed commercially, resulting in many hybrid varieties with slightly different active substances. With organic compounds such as persin, the same substance can be harmless to one species and deadly for another, although the species might be closely related. Clinical intoxications have been confirmed for Budgerigars and various parrots. The central nervous system and the heart are affected. The onset of symptoms comes usually within twelve to twenty-four hours after consumption; death occurs within two days.

Veterinarian Alan Jones relates in his book *Keeping Parrots: Understanding their care and breeding* that he has unfortunately seen the results of avocado poisoning on several occasions. When one of his clients went away, leaving her daughter to feed her six parrots, disaster struck. Each bird was given a slice of avocado, as a "treat". Several hours later the parrots showed signs of abdominal pain, with vomiting and diarrhoea. There is no antidote, but supportive treatment was given. Sadly, four of the parrots died within twenty-four hours and the other two were extremely ill for several days. Do not take any risks! This includes not leaving avocado in open containers if you have companion birds free in your home.

Pesticides on commercial fruit

Those who care for parrots should know that almost all commercially-grown cultivated fruits are sprayed with chemicals. The average conventionally grown apple is said to contain more pesticide residue than any other fruit or vegetable. Washing apples might remove some of the pesticides on the skin but research has shown that apples sprayed with pesticides retained 50% to 100% of chemical residues even after several months in cold storage and after being washed with detergent. The Environmental Working Group in the USA tested seven hundred samples of washed apples and found 48 different kinds of pesticides. In oranges they found fifteen kinds. Chilean grapes were the worst; they have caused deaths among some captive parrots.

The amount of fruit eaten by most parrots is probably not large enough for pesticide residues to be harmful under normal circumstances. Apples, raspberries and redcurrants are easy to grow – so perhaps parrot owners should grow at least some of their own to provide organically-grown fruits for part of the year.

The need for sodium, clay and minerals

In Africa, forest elephants knock down trees and make clearings to reach the mineral salts in the earth. Flocks of Grey Parrots, hundreds strong, then descend to the ground to partake of this mineral soil.

I was reminded of the importance of soil when I noticed one of my conures, who had just laid the first egg of her clutch, searching – but not for food. I immediately went and dug up a clod of earth and grass. She attacked this frantically. It was the soil that interested her. I do believe that captive parrots should have access to soil on a regular basis – soil that has not been treated with any chemical.

Insoluble grit aids in digestion and is particularly important for birds kept in cages without access to sand and dirt floors. Soluble grit is the major source of minerals in a bird's diet. It has been suggested that lack of salt (sodium) and iodine might predispose some birds to pluck themselves. Certainly iodine blocks for Budgerigars were developed decades ago as lack of this chemical element can cause ill health.

Images of big numbers of parrots at clay licks have since the 1980s graced many books and television programmes because it is such a colourful and exciting spectacle. Most of these soil banks are found in Peru, Ecuador and Bolivia, in the rainforest along the eastern base of the Andes, usually by a river. The popular theory was that parrots ingested clay to protect them against the toxins found in many seeds. Now it is widely believed that parrots are also, or only, seeking sodium.

In the Rio Cristalino rainforest reserve in Mato Grosso, Brazil, two species of *Pyrrhura* conures, Crimson-bellied and Madeira (*P.perlata* and *P.snethlagae*) seek out a small patch of earth in the forest. In the dry season this is exposed and gives them access to earth containing sodium. Ever-cautious, the conures might wait a long time high in the trees above until they are confident no predators are present. They are canopy-feeders and are nervous about descending to the ground. I saw that butterflies seek the same patch for the same reason.

Donald Brightsmith, well-known for his studies of parrots at the Tambopata research station in Peru, wrote: "In general nutrients are often in short supply in many ecosystems. In particular sodium is often cited as an important reason why animals eat soils. In fact in temperate areas most geophagy sites are referred to as salt licks. Sodium is scarce in the diets of herbivorous animals because it is found in low concentrations in most plants. Many plants actively avoid uptake of sodium. However sodium is vital for a wide variety of animal functions including maintenance of osmotic balance, nerve transmission etc. For this reason humans and other animals show such strong cravings for sodium and actively seek it out."

In Peru parrots have diets with extremely low concentrations of sodium, thus these birds eat the soils with the highest concentrations. Donald Brightsmith concluded: "The scientific evidence suggests Peruvian parrots do not eat soil for grit but they do consume soil that provides an important source of dietary sodium and helps neutralize the plant toxins in their diet."

Brightsmith believed that the annual peak in clay lick use by Scarlet Macaws was due in part to the parents' need to feed clay to their young. In eight food samples taken from the crops of chicks from twenty to thirty days old seven samples contained soil. This was true of only one of eight samples collected from chicks over fifty days of age (Brightsmith, 2004).

The need of wild parrots for sodium definitely does not mean that we should add salt to the diet of our own parrots: their diet is quite different. In particular salty foods such as salted crisps and peanuts should not be offered and might even prove poisonous in excess. In any case, these foods are high-fat and habit-forming!

Do parrots need grit?

There has been some debate regarding whether grit should be offered to captive seed-eaters. On post mortem of some birds the gizzard has been compacted with grit, with the conclusion that this was the cause of death. However, they might have been suffering from a digestive problem that caused them to consume excessive quantities. Or they had some other illness. Thus it might be advisable to remove grit from sick birds.

A post mortem photograph of the gizzard contents of a cockatoo showed that it was full of unshelled pine nuts. They had caused a compaction. The breeder believed that, in the absence of grit, the cockatoos had swallowed the small nuts whole. He commented: "A £600 cockatoo died for the lack of 10p worth of grit."

I believe that grit is beneficial to the well-being of most seed-eaters. Birds fed mainly on seed and kept without access to fresh soil or green plants with roots and soil, or not given mineral supplements, will be deficient in trace elements. A deficiency of manganese, for example, can result in dead-in-shell and premature ageing. The major elements, such as calcium, phosphorus, sodium and chlorine, are found in grit. Sodium assists digestion. Grit costs so little my suggestion would be to provide a small container of it or to sprinkle it on the aviary floor, and let the parrots decide if they need it.

In the encyclopaedic work, *Avian Medicine, Principles and Application,* it states: "Grit is not required in the normal, healthy psittacine or passerine bird. Grit, defined as a granular, dense, insoluble mineral material (generally granite or quartz) is required in birds that consume whole, intact seeds" (Ritchie, Harrison and Harrison, 1994). Birds fed formulated diets (pellets and extruded foods) are unlikely to need grit, either soluble or insoluble. This is presumably because these soft foods do not need to be ground up in the gizzard and because the necessary minerals have been added to the food.

When I asked Dr Stacey Gelis from Melbourne for his views as an aviculturist and veterinarian, he told me: "Grit is not essential in parrots, but that is not to say that they do not benefit from it or, indeed, do not like it. I think that birds benefit from eating soluble grits, especially ground-feeding species. One point worth mentioning is that birds which do not dehusk their seed [pigeons, chickens, etc] often have thinner walls to their crops if not fed grit compared to those that are offered grit."

Crop contents of many wild parrots examined for scientific studies have been found to contain grit.

In Australia these include Red-cheeked Parrots *(Geoffroyus),* various Rosellas, and Galahs and Greater Sulphur-crested Cockatoos. Three of the latter collected in New South Wales had swallowed small quartz pebbles. The crop contents of nine Many-coloured Parrots *(Psephotus varius)* collected by scientists included fine grit, sand and charcoal. Fine grit and sand are swallowed by various species of *Neophema* parakeets. In Australia many parrots and cockatoos are killed on roads while collecting grit. They are not, of course, species that swallow grain whole.

Even lorikeets have been seen to pick up grit and swallow it. As they are mainly nectar- and pollen-eating birds, grit is not used in the gizzard (also called the ventriculus) to grind food, as it is in seedeaters. The gizzard is poorly developed in lories. I suspect that many birds seek grit for the minerals they contain.

Different kinds of grit are packaged by bird food manufacturers. Mineral grit consists of a variety of small stones such as limestone and oystershell, which are made of calcium carbonate, also quartz and charcoal (soluble). The bird might select the kind it needs and leave the rest. Thus the grit container should be completely emptied and refilled at intervals.

Calcium carbonate is digested by acids in the proventriculus and is therefore of no use in grinding down food, but is a good source of calcium. The proventriculus is the first part of a bird's stomach, where digestive enzymes act upon the food before it goes to the gizzard.

Cuttlefish bone is commonly fed to parrots for its calcium content. Why offer it when one can use a variety of calcium supplements specially formulated for birds? Calcium cannot be absorbed without Vitamin D in the diet or unless the bird has access to sunlight. In theory, then, a bird fed mainly on seed or nothing but seed and kept indoors would gain little benefit from cuttlefish bone as it would be unable to absorb the calcium it contains. However I have observed the almost desperate, way in which some females consume it before they lay. They know instinctively when they need it. However, some of my male lories consume cuttlefish bone with equal enthusiasm. I therefore suspect that there is a salt content that is also attractive to parrots. I do believe that calcium should be available to females in three forms: grit, cuttlefish bone and a calcium supplement containing the vital Vitamin D.

Certain enzymes (complex proteins produced by cells that promote specific biochemical reactions) are important to parrots. Enzymes cannot function properly if the diet is deficient in certain minerals or trace elements in minute proportions. These include zinc, iron, iodine, manganese, sulphur, selenium, cobalt and molybdenum – metallic and non-metallic elements not found in foods, or found in infinitesimal quantities. They are probably available in typical bird grit mixtures, also in powdered mineral supplements.

PART VII.
STOP AND THINK...

Now extinct. Gone forever...One Glaucous Macaw was the pet of a wealthy lady in England in the 1920s.

Sketch by Paul Staveley © Rosemary Low

37. CAPTIVE BREEDING IS NOT CONSERVATION

Breeders should be realistic and accept that captive breeding is not conservation. They might argue that keeping a species alive in aviaries is preferable to them becoming extinct. I agree – up to a point. The problem is that breeders are purposely or inadvertently selecting for features that result in generations diverging from the species in the wild. Cage-breeding often results in small and degenerate specimens. Too many parrots are reared in accommodation so small they are unable to fly. Whether or not they could survive in the wild would be immaterial in most endangered species because loss of habitat is the reason for their decline.

In-breeding is another problem that results in degenerate young. Breeders selling young "pairs" without admitting that they are siblings is totally unacceptable.

The emphasis on producing mutations in countless parrot species means that these birds are losing their resemblance to wild birds, as the original wild type, genetically pure, becomes harder to find. The same applies to the production of hybrids.

We have only to look at how far the exhibition type Budgerigar has diverged from the beautiful, elegant little wild birds to see what captive breeding can do. Release an exhibition Budgerigar into the wild and its reduced powers of flight would probably mean it could not even reach the middle branches of a *Eucalyptus microtheca* tree. It is true that the Budgerigar is an extreme example – but no other parrot has been bred in captivity on a regular basis since the 1850s. It should be looked on as a warning of what generations of captive breeding can do.

Over-production of parrots and parrot rescue centres unable to accept more birds are unfortunately now the norm. Jamie Gilardi, Director of the World Parrot Trust, wrote: "Are we kidding ourselves when we say that we're breeding birds 'for conservation'? It is highly unlikely that parrots now in private hands will ever be of direct conservation value to the species in the wild. We should all think long and hard about why we own birds and, if we chose to breed them, why we've made that choice. There are legitimate reasons to breed parrots in captivity: conservation simply is not one of them."

The title of this chapter will not please some aviculturists. In fact it should have been entitled "Captive Breeding by Aviculturists is not Conservation." I am relating this statement only to parrots and, of course, point out below that there are a handful of instances where government-approved conservation projects have included captive breeding and release. These are all in the country of origin and close to the natural habitat.

Breeding endangered species

I was completing this book by working on the index when I received an e-mail that made me realise I had not touched on this subject. I need to do so because there is much misunderstanding as my correspondent indicates.

Mr X *I would like to help some of the endangered species of parrot, by breeding and rearing in captivity. I have approximately a quarter of an acre and I plan to construct the largest and best aviaries. I think that I can offer the birds a great enriched environment. They will have large flights, partly covered. Can you advise on how I would go about obtaining the required birds?*

RL *Just one question before I proceed: do you already have experience in breeding parrots?*

Mr X *I have no experience with parrots. I have bred Budgerigars and also hand-reared them. I have been reading your books and feeding my brain with as much information as I can absorb and act upon. I am extremely confident that I will be able to cope, but I don't take it lightly. I am excited about it and looking forward to increasing a species. I would like advice on reintroduction to the wild. Is this possible?*

RL *I am afraid that reintroduction of parrots bred by aviculturists to the wild is a myth. The only cases where captive-bred parrots have been released into their natural habitat are in in situ breeding programmes, such as the Puerto Rican Parrot (sponsored by government agencies), the Echo Parakeet in Mauritius (various international conservation organisations) and Scarlet Macaw in Costa Rica (under strict supervision of the government, for disease control, etc).*

The disease risk from captive-bred birds is too high to release them except in a few programmes where extensive (and expensive) tests for various diseases are carried out. These days viral diseases in captive parrots are a big problem and could decimate wild populations if diseased birds were released.

Puerto Rican Parrot in a US Government breeding facility on Puerto Rico.

It is therefore unlikely that private aviculturists would be involved in such programmes but one bird park is currently taking part. After much negotiation with government agencies, in 2013 six Blue-throated Macaws (Ara glaucogularis) *bred at Paradise Park in Cornwall, UK, home of the World Parrot Trust, were sent to Bolivia. The hope is to release them eventually. Unlike the situation for most endangered parrot species where suitable habitat for release no longer exists, the Blue-throated Macaw was trapped almost out of existence so there are areas where it could be reintroduced and might not come into contact with wild macaws.*

You have a good grounding in breeding Budgerigars but this is very different to breeding the larger parrots. My advice would be to breed parrots for enjoyment only and to start with a couple of pairs of the more common species to gain experience. Sorry to dampen your enthusiasm but you need to know the facts.

My concern was for rare parrots that might fall into inexperienced hands and I was rather apprehensive about how this advice might be taken. I did not want to offend but I did want to be totally honest. To encourage a beginner to start with endangered species would be irresponsible.

Mr X *Obviously I am very disappointed. I thought that I could make a small difference in parrot species' survival. I will take your advice and start slowly, with some common parrots. If I cannot help with a hands-on approach, I am considering funding one or more of the in-situ breeding programmes. Cannot tell you enough, thank you for your time and help.*

RL *I am so glad you were not offended by my answer and that furthermore you are considering donating to parrot conservation. This is an extremely sensible and valuable approach and I admire you for it.*

The mention above of sending Blue-throated Macaws to Bolivia needs further explanation. In 1992 Paradise Park received six young captive-bred birds. Curator David Woolcock wrote: "The character and charm of these wonderful birds immediately endeared them to staff and visitors alike."

Difficulties in "repatriation"

The dream of Paradise Park was always to return young to Bolivia, so the birds were managed to minimise any risks that might compromise that ambition. However, it was not until February 2004 that the first egg was laid. When a number of young had been reared, it took until August 2012 (twenty years after the first macaws were received) to gain permission from the Bolivian Government for the World Parrot Trust to send them to Bolivia. Seven macaws were chosen, six of which had been housed together as a group.

To indicate how complicated the process was, I will relate the difficulties that ensued. The Bolivian ministry required health certification – but no birds had ever previously been sent from the UK to Bolivia so Paradise Park had to create a document and have it approved by the Bolivian ministry. The birds then had to go through a 60-day quarantine period in the UK during which many health tests took place.

The plan had been to fly the birds to Miami from Heathrow and then on to Santa Cruz, Bolivia. Due to logistical reasons this was impossible. In order to achieve a route two airlines had to work together. The next problem was that the new route was via Madrid. As the birds would be leaving the EU there an additional health certificate was acquired for the European leg of the journey, thus delaying departure.

The planned date of January 23 had to be changed to February 20, flights re-booked and veterinary visits to sign health certificates rearranged. This was due to access roads to the Bolivian site being flooded and impassable. Then there was another obstacle. The Heathrow to Madrid aircraft had tie-down spots for only six crates, so only six macaws could be sent. But that was not the end of the story. An impending handling agent strike in Madrid was scheduled for February 18 to 22. The chances were that once they got to Madrid the macaws would be stuck there until the strike was over. The only solution was to reschedule the departure for February 27 which, of course, meant re-booking all flights, vet visits, etc (Woolcock, 2013). After travelling for fifty hours (delayed by snow in Madrid), the macaws finally reached their destination, after being accompanied from Madrid.

This story will help to explain why, in the unlikely event of a private aviculturist obtaining permission to send parrots back to the country of origin, few would be inclined to try! And few would have the patience to carry it through!

Habitat loss and restoration

It must be emphasised that many parrots are now endangered due to loss of habitat and there are no locations where they could be released. The need for knowledge regarding this point was brought to my attention (again as I was preparing this chapter) by a lady in Canada who had three Swift Parrots. When she learned that this was an endangered species she contacted a parrot rescue centre to ask if she could donate them to an organisation breeding endangered parrots. The rescue centre asked for my suggestions.

I pointed out that captive breeding of the Swift Parrot in North America would not assist its survival in Australia. In any case that country does not permit the importation of parrots – and even if it did, releasing captive-bred birds would be pointless. Its numbers could increase in the wild only if its habitat is protected, improved and expanded. Habitat loss and the more recent discovery of sugar gliders predating females and young in the nest are driving it towards extinction. I suggested that the Swift Parrots should remain in their current location – as they were with someone who obviously cared a lot about them.

These days parrot conservation focuses on protecting and improving habitat and educating local people about the reasons to conserve parrots. Providing nesting sites (because so many large trees have been destroyed) is increasingly being employed as an effective – and even rapid – method of increasing numbers.

South and Central America have been the focus for most of these initiatives. In Asia and the Pacific region few parrot conservation projects exist. It seems likely that some parrot species there are in serious trouble – but little is known about most populations.

There is much misunderstanding and misinformation about captive breeding saving parrots from extinction. Only a few days before I received the above e-mails, a newspaper in the UK published an opinion which read:

The survival of Spix macaws has been dependent on captive breeding. Lear's and glaucous macaws fall into the same category. The California Condor was saved by breeding techniques developed by aviculturists, not academics.

I felt obliged to reply to comment on the inaccuracies and my comments were published as follows:

Your contributor is right about the California Condor but not about the macaws. Glaucous is extinct in the wild, probably since the 1930s when there were two reports that it had survived in the Corrientes region of Argentina. There were reports in Paraná, Brazil, in the early 1960s – but unsubstantiated.

The well-known British aviculturist Sydney Porter, reported seeing one, aged at least 45, in Buenos

Young Spix's Macaw bred at Loro Parque, Tenerife.

Aires Zoo in 1938. That individual could have been the last glaucous macaw on the planet. A few isolated birds were known in zoos, including London and Amsterdam, and there was one bird in a private collection in New York in 1925. Captive breeding never occurred.

Now Spix's macaw. True its survival has been dependent on captive breeding, but it was actually trapped to extinction – except for the very last one. In 1987 the female of the last pair was torn from the nest by trappers. Her mate managed to evade capture and fought his way to freedom and fame as the last wild Spix's macaw on the planet.

Two collections, including Loro Parque, Tenerife, have been very successful in breeding this species. Yet we must remember that no Spix's Macaw was ever exported legally. It was wealthy aviculturists who caused its extinction in the wild as they paid trappers huge sums.*

Lear's macaw has been bred with great success in only one location: again, Loro Parque. They have achieved astounding results with a small number of birds. However, this species, too, was trapped almost to extinction. It was not saved by captive breeding but by field work, including round-the-clock guards on nests, supported by several organisations, including Fundacion Loro Parque (LPF).

Last year I was in Brazil and observed first-hand the work of a number of extraordinary people, some of whom are virtually dedicating their lives to saving this macaw. Its survival is not "dependent on captive breeding" but on extremely dedicated field workers, conservationists and parrot conservation organisations.

I would urge everyone interested in parrot conservation to support this work by donating through such organisations as LPF and Parrots International. The Parrot Society UK has recently donated £5000. But so much more is needed if this charismatic macaw is to survive over the long term.

---

*except in later years to Loro Parque with the permission of the Brazilian Government, whose property they remained.

The sad illegal capture of the last wild Spix's Macaw on the planet.

It is a bizarre realisation that in 1913 in a house in Cambridge a maid took a macaw out of its cage and cradled it in her arms like a baby. It was a gentle and beloved pet of a Mrs Anningson. The maid was holding one of the most precious birds in the world – but she did not know it. The macaw was a Glaucous.

There is no doubt about this. Wesley Page, editor of the magazine *Bird Notes,* recorded his visits in 1908 and 1913. It was ironic that on the first occasion he congratulated Mrs Anningson on the acquisition of such a rare creature as a Hyacinthine Macaw – but noted "more rare if anything than the preceding which it closely resembles" was a Glaucous Macaw. His description of the plumage left no doubt that this was so. He had not misidentified a Lear's Macaw.

As I recalled this story, which I told previously in *A Century of Parrots* (Low, 2006), I could not help wondering if today, someone, somewhere, is cradling a pet parrot of a species which will be extinct one century from now. I fear the answer is yes.

38. THE WRONGS WE HAVE HEAPED ON PARROTS

The trade in wild-caught parrots has had a terrible impact on parrot populations. Combined with devastating deforestation throughout the tropics, especially since the second half of the 20th century, the result is that nearly one third of species are threatened with extinction.

Many people believe that capturing wild-caught parrots has diminished. In some species this is the case. Take, however, the trade in the world's most famous "talking" bird: the African Grey Parrot. In 2011 breeders in South Africa imported more than 5,000 wild-caught Greys from the Democratic Republic of Congo (DRC), as well as from other central African states. They exported nearly 25,000 Greys to many countries worldwide. These reputedly included captive-bred birds in addition to wild-caught birds laundered through breeders and exported as captive-bred (Bradshaw and Engebretson, 2013). (See also **21. The Basics:** Carbon Monoxide Poisoning).

Although the European Union banned the importation of wild-caught parrots in 2006, CITES continued to support export from the DRC, from the Congo and from Cameroon to South Africa, South-east Asia and the Persian Gulf. Chinese business initiatives in Africa seem to have resulted in a rising interest in Grey Parrots in China where the potential demand is huge. It always seems to be that when one market closes another opens. This leads me to wonder if the capture of parrots will cease before some species are completely exterminated by trade.

Worldwide there are millions of wild-caught parrots in captivity. Many of them do not make suitable companion birds because they were captured when adult and retain their fear of man. They lead sad lives, perhaps rehomed many times and bought by people who have no idea of the tortures they have suffered, especially when first captured. They have long memories and many are afraid of hands, men and sticks.

In close confinement they are fearful and aggressive and, in the case of Grey Parrots, growl at the approach of a human. Such birds are suited only to life in a large aviary where they have no need to interact closely with people. You can see the fear in their eyes, and striking out at people who come near is their only defence.

As I was working on this book I received an e-mail from someone who had bought a female wild-caught Grey which was kept in an outdoor aviary with a semi-tame Grey. She described the female as very frightened of people and said she wanted it to be added to the list of parrots waiting to be returned to the wild. It is good to know there are people who are concerned enough about the unhappy state of mind of a wild-caught parrot. Sadly, it is but a pipe-dream that such birds could go back to the wild. As described in the previous chapter, this is not practical – and never will be.

The status of the Grey Parrot, previously considered to be very common, has been amended to Vulnerable, mainly due to excessive and unsustainable trade. The IUCN category Vulnerable means a species is facing a 10% risk of extinction within the next 100 years. Most people consider the Grey Parrot to be a common species. Its extinction in the wild is unthinkable. Yet so was that of the Carolina Parakeet *(Conuropsis carolinensis)* described as so numerous that flocks covered the fields of grain "so entirely, that they present to the eye the same effect as if a brilliantly coloured carpet had been thrown over them." Thus wrote the artist John James Audubon in 1831.

Destroying the woodlands in which it fed and shooting the parakeet as a crop pest contributed to its extinction. Now, nearly 200 years later, some parrot species have been brought back from the edge by the combined efforts of dedicated field workers, scientists and people who contribute financially to conservation. There will never be enough resources to save all the threatened species in a world where their habitat is diminishing at a frighteningly rapid speed, due to over-population of the human race and their demands for land, minerals, oil and water.

In 1998 Stuart Taylor received reports that Brown-headed Parrots were being traded in Mozambique. Through the Research Centre for African Parrot Conservation he applied to an official in that country to investigate this. The proposal included catching and ringing parrots to provide future data. The official looked through the proposal with an obvious lack of interest. Then he reached the part about catching parrots and asked: "How many can I have? I would like six." It was explained that the parrots would be released immediately. The official did not make further contact.

Sadly, conservationists and researchers trying to save parrots face innumerable difficulties. Human attitude is only one of them. The – to me – terrifying human population growth of the last century or so has resulted in loss of habitat on a scale and geographical extent that means parrot extinctions within the next few decades are inevitable. Already too many species are hanging by a thread on to existence. As mentioned in the Introduction, 28% of all the 350 parrot species are threatened with extinction (as of 2014). I make no apology for repeating this fact here.

Parrots evolved eighty to ninety million years ago but the oldest fossils of modern parrot families date back about twenty million years. In the space of a few decades we are making a good job of ensuring that many will soon be lost forever, as were the Glaucous Macaw and the Carolina Parakeet in the 20th century.

What can we do?

The message is simple – and highly unpopular. Human over-population and consumerism is driving the destruction of the earth's habitats and resources and wildlife. Not until humans learn to limit their families to one or two children and to show respect for nature and all the wonderful creatures on the planet, will the tide of destruction start to turn.

REFERENCES CITED

Albertani, F.P., C.Y.Miyaki and A. Wanjtal, 1997. Extra-pair paternity in the Golden Conure *(Guaruba guarouba)* detected in captivity, *Ararajuba* 5 (2): 135-139.

Albornoz, M.B. and A. Fernández-Badillo, 1994. Psitácidos (Aves: Psittaciformes) plagas de cultivos en el valle del río Güey, estado Aragua, Venezuela, *Rev. Fac. Agron.* (Maracay) 20: 123-132.

Arnold, K.E., 2002. Fluorescent signalling in parrots, *Science,* Jan. 4, 295, No. 5552: 92 (published on-line).

Auersperg, A.M.I., B. Szabo, A.M. P. von Bayern and A. Kacelnik, 2012. Spontaneous innovation in tool manufacture and use in a Goffin's cockatoo. *Current Biology,* 22 (21) online page 6, November.

Barreira, A.S., M.G. Lagorio, D.A. Litjmaer, S.C. Lougheed and P.L.Tubaro, 2012. Fluorescent and ultraviolet sexual dichromatism in the blue-winged parrotlet, *Jnl of Zoology:* 135-142.

Beckers, G.J., B.S. Nelson and R.A. Suthers, 2004. Vocal-tract filtering by lingual articulation in a parrot, *Current Biology,* 14: 1592-7.

Beissinger, S.R., 2008. Long-term studies of the Green-rumped Parrotlet *(Forpus passerinus)* in Venezuela: hatching asynchrony, social system and population structure, *Ornitologia Neotropical* 19 (suppl): 73-83.

Berg, K.S., S. Delgado, K. Cortopassi, S. Beissinger and J. Bradbury, 2011. Published online 13/7/2011, Vertical transmission of learned signatures in a wild parrot, *Proc. Royal Society.*

Berlin, K.E. and A.B. Clark, 1998. Embryonic Calls as Care-soliciting Signals in Budgerigars, *Melopsittacus undulatus, Ethology,* 104, 531-544.

Birkhead, T., 2008, The Wisdom of Birds, Bloomsbury.

-----2012. *Bird Sense,* Bloomsbury, London.

Boles, W., 1991. Glowing Parrots, *Birds International* 3: 76-79.

Bollen, A. and L. Van Elsacker, 2004. The feeding ecology of the Lesser Vasa Parrot *(Coracopsis nigra)* in south-eastern Madagascar, *Ostrich* 75 (3): 141-146.

Boussekey, M., J. Saint-Pie and O. Morvan, 1991. Observations on a Population of Red-fronted Macaws *Ara rubrogenys* in the Río Caine Valley, central Bolivia, *Bird Conservation International,* 1: 335-350.

Bradbury, J. and T. Balsby, 2006. The mystery of mimicry, *PsittaScene,* 18 (3): 8-11.

Bradshaw, G. A. and M. Engebretson, 2013. Parrot Breeding and Keeping: the impact of capture and captivity, Policy Paper, Animals and Society Institute, USA.

Bradshaw, G.A., J.P. Yenkosky and E. McCarthy, 2009. Avian Affective Dysregulation: Psychiatric Models and Treatment for Parrots in Captivity, *Proceedings of 30th Annual Association of Avian Veterinarians Conference.*

Brightsmith, D., 2004. Effects of Diet, Migration, and Breeding on Clay Lick Use by Parrots in Southeastern Peru, *American Federation of Aviculture 2004 Symposium Proceedings.*

Burbridge, A.H., 2008. Little and Long-billed Corellas Learning to Use A New Food Source, the Seeds of Marri, *Australian Field Ornithology,* 25:136-139.

Burger, J. and M. Gochfeld, 2003. Parrot behavior at a Rio Manu (Peru) clay lick: temporal patterns, associations, and antipredator responses. Published online: 1 October 2003, Springer-Verlag and ISPA.

Carter, M., 1996. Nesting Rosellas (*Platycercus* spp.): Innovative Site Selection and Notes On Repeat Breeding and Other Behaviour, *Australian Bird Watcher,* 16: 344-348.

Clark, P., 2002. Behavior as a reflection of breeding and rearing practices, *Companion Parrot Quarterly,* 57: 92-95.

Cockle, K. L. and A. Bodrati, 2011. Vinaceous Parrot *(Amazona vinacea),* Neotropical Birds Online (T. S. Schulenberg, Editor), Ithaca: Cornell Lab of Ornithology.

Cravens, E., 2003. The Pushy Parrot, *Companion Parrot Quarterly,* 60: 80-83.

Csaky, K.K., 2014. My Hahn's Macaw is attacking everyone! *Parrots,* February: 28-30.

Dawson, W.R. and C.D. Fisher, 1982. Observations on the temperature regulation and water economy of the Galah *(Cacatua roseicapilla), Comparative Biochemistry,* 72A, 1-10.

Dodman, N.H., A. Moon-Fanelli, P.A. Mertens et al, 1997. Veterinary models of OCD *in Obsessive-Compulsive Disorders* (E. Hollander and D. Stein, eds), Marcel Dekkar, New York: 99-143.

Forbes, N.A., 2014. Meeting the modern welfare requirements for pet parrots or time for some 'navel gazing'? *Parrots,* June: 36- 38.

Forshaw, J.M., 1998. Fig Parrots in the wild, *Cage & Aviary Birds,* November 28: 5.

----- 2002. *Australian Parrots* (third revised ed.), Alexander Editions, Queensland.

Friis, A., 2014. Discovering Bourke's Parakeets in the wild, *Parrots,* January: 28-30.

Greene, T.C. and J.R. Fraser, 1998. Sex Ratio of North Island Kaka *(Nestor meridionalis septentrionalis),* Waihah Ecological Area, Pureora Forest Park, *New Zealand Journal of Ecology,* 22: 11-16.

Groszmann, R., 2002. Breeding and Parenting Behaviour in a Pair of Greenwinged Macaws, *Companion Parrot Quarterly,* 57: 2-14.

Gsell, A.C., J.C. Hagelin and D.H. Brunton, 2012. Olfactory sensitivity in Kea and Kaka, *Emu,* 112 (1).

Halaouate, M., 2014. Assessing the Red-and-blue Lories on Karakelang Island, *Parrots* May: 36-38.

Hausmann, F., K. E. Arnold, N. J. Marshall and I. P. F. Owens, 2003. Ultraviolet signals in birds are special, *Proc. Royal Soc. of London,* 270: 61-67.

Heinsohn, R., 2012. True Colours, *PsittaScene,* 24 (2): 3-7.

Hile, A.G. and G.F. Striedter, 2000. Call convergence within groups of female budgerigars *(Melopsittacus undulatus), Ethology* 106: 1105-1114.

Hirst, L., 2014. Through the eyes of a bird, *Parrots,* January: 31.

Holyoak, D.T., 1972. Adaptive Significance of Bill Shape in the Palm Cockatoo *(Probosciger aterrimus), Avicultural Magazine* 78 (3): 99-100.

Horstman, E., 2014. The Rescue, *PsittaScene,* Summer: 10-11.

Krasheninnikova, A., S. Bräger and R. Wanker, 2013. Means-end comprehension in four parrot species; explained by social complexity, *Animal Cognition,* 16 (5): 755-764.

Kyle, T., 2009. Sun Conures rising, *PsittaScene,* November: 3-5.

Low, R., 1992. Breeding Pesquet's Parrot, *Avicultural Magazine,* 98 (4): 163-172.

----- 1998. *Encyclopedia of the Lories,* Hancock House, Canada/USA.

----- 2006. *A Century of Parrots,* Insignis Publications, Mansfield.

----- 2009. *Go West for Parrots!* Insignis Publications, Mansfield.

----- 2013. *Pyrrhura Parakeets (Conures): Aviculture, Natural History, Conservation,* Insignis Publications, Mansfield.

Lumholtz, C., 1902, *Among Cannibals,* Charles Scribner's & Sons, New York.

McKendry, J., 2013. Flight Status, *PsittaScene,* Autumn: 16-19.

Martin, S., 2012. Blue-throated Macaw, *PsittaScene* August: 7.

Masello, J.F., A. Sramkora, P. Quillfeldt, J.T. Epplen and T. Lubjuhn, 2002. Genetic monogamy in burrowing parrots *Cyanoliseus patagonus? J. of Avian Biology* 33: 99-103.

Masin, S., R. Massa and L. Bottoni, 2004. Evidence of tutoring in the development of subsong in newly-fledged Meyer's Parrots *Poicephalus meyeri. Annals of the Brazilian Academy of Sciences,* 76 (2): 231-236.

Metz, S., 2002. Inside the minds of parrots, *Companion Parrot Quarterly,* 57: 14-18.

Moura, L.N. de, J. M. Vielliard and L.M. Da Silva, 2010. Seasonal fluctuation of the Orange-winged Amazon at a roosting site in Amazonia, *Wilson Journal of Ornithology.*

Moura, L.N., M.L. Silva, M.M.F. Garotti, A.L.F. Rodrigues, A.C. Santos and I.F. Ribeiro, 2014. Gestural communication in a new world parrot, *Behavioural Processes.*

O'Brien, S., 2008. *The Story of a Remarkable Owl,* Constable, London.

Perrin, M., 2012. *Parrots of Africa, Madagascar and the Mascarene Islands,* Wits University Press, Johannesburg.

Pizo, M.A., C.I. Donatti, N.M.R. Guedes and M. Galetti, 2008. Conservation puzzle: endangered Hyacinth Macaw depends on its nest predator for reproduction, *Biological Conservation,* 141 (3): 792-796.

Ritchie, B.W., G.J. Harrison and L.R. Harrison, 1994. *Avian Medicine: Principles and Application,* Wingers Publishing Inc., Florida.

Roper, T.J., 2003. Olfactory discrimination in Yellow-backed Chattering lories *Lorius garrulus flavopalliatus:* first demonstration of olfaction in Psittaciformes, *Ibis,* 145: 689-691.

Rowley, I., 1990. *Behavioural Ecology of the Galah,* Surrey Beatty & Sons, Australia.

Saidenburg, A.M., N.M.R. Guedes, G.H.F. Seixas, M.da Costa Allgayer, E. Pacifico de Assis, L.F. Silveira, P.A. Melville and N. Roberti Benites, 2012. A survey for *Escherichia coli* Virulence Factors in Asymptomatic Free-Ranging Parrots. Published online 2012 July 10, *ISRN Veterinary Science.*

Santos, I.C.O., B. Elward and J.T. Lumeij, 2009. Sexual Dichromatism in the Blue-fronted Amazon Parrot *(Amazona aestiva)* revealed by Multiple-angle Spectrometry, *Journal of Avian Medicine and Surgery.*

Saunders, D.A., 1977. Red-tailed Black Cockatoo breeding twice a year in the south-west of Western Australia, *Emu,* 77: 207-110.

----- 1983. Vocal repertoire and individual vocal recognition in the short-billed white-tailed black cockatoo, *Calyptorhynchus funereus latirostris* Carnaby. *Australian Wildlife Research* 10: 527-536.

Schmutz, 1977. Die Vogel der Manggarai (Flores), 2, Niederschrift. Unpublished.

Smith, G.T., 1991. Breeding ecology of the Western Long-billed Corella, *Cacatua pastinator pastinator, Australian Wildlife Research,* 18: 91-110.

Snyder, N.F.R., J.W. Wiley and C.B. Kepler, 1987. *The Parrots of Luquillo: Natural History and Conservation of the Puerto Rican Parrot,* Western Foundation of Vertebrate Zoology.

Speed, N., 2003, Parrots Kept for Breeding, *Companion Parrot Quarterly,* 60: 26-30.

Tamungang, S.A. and S.S.Ajayi, 2003. Food diversity of the Grey Parrot *Psittacus erithacus* in Korup National Park and its Support Zone, Cameroon. *Bulletin African Bird Club,* 10 (1): 33-36.

Viader, X., 2010. Grey Parrot management and breeding, *Proceedings VII International Parrot Convention,* Loro Parque, Tenerife: 121-130.

Walfield, P., 2003. Teaching little Louie to fly, *Companion Parrot Quarterly,* 60: 32-33.

Walker, J.S. and M. Seroji, 2000. Nesting behaviour of the Yellowish-breasted Racquet-tail *Prioniturus flavicans, Forktail,* 16: 61-63.

Woolcock, D., 2013. Dream Come True, *PsittaScene,* 25 (2): 6-7.

INDEX – SPECIES only

To look up the scientific name of a parrot species, refer to the page on which it is first mentioned. Numbers in **bold** refer to an illustration of or an illustration relating to that species.

NOTES

ALSO by Rosemary Low

Parrots and Finches: Healthy Nutrition

163 pages, 47 colour photos.

Very occasionally an avicultural book comes along that is an interesting read and breaks new ground. In my opinion this book is one. It is highly important as a reference tool to all birdkeepers and I would urge everyone to buy, beg or borrow (but not necessarily steal) a copy. This will become one of the "classics", referred to for many years to come.

– Darren Sefton, *Foreign Birds,* Winter 2012. *

Go West for Parrots!

262 pages, numerous photographs and maps.

For those of you who think that Rosemary Low only writes books about parrots – think again. *Go West for Parrots!* – is a fascinating insight into her travels in South America over the past 35 years and the vast array of species and people she encountered. It is as much a travel log as a bird book, equally treasured by the traveller as by the naturalist. With its stunning cover shot of Hyacinthine Macaws flying over the Pantanal to its charming hand-drawn maps, this book is one which you will find difficult to put down. It deserves a much wider audience than the avicultural and natural history market.

– David Woolcock, Curator, Paradise Park, Cornwall. **

Published by Insignis Publications:
***£15.45, ** £14.55 post paid in UK.**